'This book substantially advances our understanding of human-place interactions and the nature of the human-place relationship in tourism research with implications for addressing various challenges facing tourism destinations in the world today. It should be required reading for social scientists, tourism practitioners and policy makers'.

Larry Dwyer, *Visiting Research Professor, Business School, University of Technology, Sydney*

'With great insights on topophilia and place authenticity studies, this refreshing book walks the reader through theoretical discussions on sense of place and place attachment and their relevance to humanistic geography and tourism, together with interesting examples and cases drawn from various cultures and contexts'.

Bihu Wu, *Professor and Director, International Center for Recreation and Tourism Research, Peking University*

Sense of Place and Place Attachment in Tourism

Place is integral to tourism. In tourism, almost all issues can ultimately be traced back to human–place interactions and human–place relationships. Sense of place, also referred to as place attachment, topophilia, and community sentiment, has received significant attention in tourism studies because it both contributes to, and is affected by, tourism.

This book, written by notable authors in the field, examines sense of place and place attachment in terms of a typology of sense of place/place attachment that includes genealogical/historical, narrative/cultural, economic, ideological, cosmological, and dynamic elements. Dimensions of place attachment such as place identity, place dependence, and affective attachment are discussed as well as place marketing, place making, and destination management.

Complete with a range of illustrative international cases and examples ranging from Santa Claus to the importance of place in indigenous and traditional cultures, this book represents a substantial addition to knowledge on the inseparable relationship between tourism and place and will be of great interest to all upper-level students and researchers of Tourism.

Ning Chris Chen is a Senior Lecturer in Marketing at the Department of Management, Marketing, and Entrepreneurship, University of Canterbury, New Zealand. His research interests are in sense of place, place attachment, resident/tourist psychology, and behaviour, sports marketing, and word-of-mouth behaviours.

C. Michael Hall is Professor, Department of Management, Marketing and Entrepreneurship, University of Canterbury, New Zealand; Docent in Geography, University of Oulu, Finland; Visiting Professor, Linnaeus University, Kalmar, Sweden; and Guest Professor in Service Management and Service Studies, Lund University, Helsingborg, Sweden. His interests include tourism, regional development, heritage, food, sustainability, and global environmental change.

Girish Prayag is a Professor of Marketing at the UC Business School, University of Canterbury, New Zealand. His research interests are related to consumption experiences in the tourism, hospitality, and event industries. He has published extensively on tourist emotions and place attachment.

Contemporary Geographies of Leisure, Tourism and Mobility

Series Editor: C. Michael Hall,
Professor at the Department of Management,
College of Business and Economics, University of Canterbury, Christchurch, New Zealand

The aim of this series is to explore and communicate the intersections and relationships between leisure, tourism, and human mobility within the social sciences. It will incorporate both traditional and new perspectives on leisure and tourism from contemporary geography, e.g. notions of identity, representation, and culture, while also providing for perspectives from cognate areas such as anthropology, cultural studies, gastronomy and food studies, marketing, policy studies and political economy, regional and urban planning, and sociology, within the development of an integrated field of leisure and tourism studies.

Also, increasingly, tourism and leisure are regarded as steps in a continuum of human mobility. Inclusion of mobility in the series offers the prospect to examine the relationship between tourism and migration, the sojourner, educational travel, second home, and retirement travel phenomena.

The series comprises two strands:

Contemporary Geographies of Leisure, Tourism and Mobility aims to address the needs of students and academics, and the titles will be published simultaneously in hardback and paperback.

Routledge Studies in Contemporary Geographies of Leisure, Tourism and Mobility is a forum for innovative new research intended for research students and academics, and the titles will initially be available in hardback only. Titles include:

Tourism in Iran
Challenges, Development and Issues
Edited by Siamak Seyfi and C. Michael Hall

Tourism Policy and Planning Implementation
Issues and Challenges
Edited by Konstantinos Andriotis, Dimitrios Stylidis and Adi Weidenfeld

Tourism and Hospitality in Conflict-Ridden Destinations
Edited by Rami Isaac, Erdinç Çakmak and Richard Butler

For more information about this series, please visit: www.routledge.com/Contemporary-Geographies-of-Leisure-Tourism-and-Mobility/book-series/SE0522

Sense of Place and Place Attachment in Tourism

Ning Chris Chen, C. Michael Hall and Girish Prayag

Routledge
Taylor & Francis Group
LONDON AND NEW YORK

First published 2021
by Routledge
2 Park Square, Milton Park, Abingdon, Oxon OX14 4RN

and by Routledge
605 Third Avenue, New York, NY 10158

Routledge is an imprint of the Taylor & Francis Group, an informa business

British Library Cataloguing-in-Publication Data
A catalogue record for this book is available from the British Library

Library of Congress Cataloging-in-Publication Data
Names: Chen, Ning Chris, author. | Hall, C. Michael, author. | Prayag, Girish, author.
Title: Sense of place and place attachment in tourism / Ning Chris Chen, C. Michael Hall, Girish Prayag.
Description: Abingdon, Oxon ; New York, NY : Routledge, 2021. | Series: Routledge focus on tourism and hospitality | Includes bibliographical references and index.
Identifiers: LCCN 2020054006 (print) | LCCN 2020054007 (ebook)
Subjects: LCSH: Tourism--Psychological aspects. | Place attachment.
Classification: LCC G156.5.P79 C54 2021 (print) | LCC G156.5.P79 (ebook) | DDC 338.4/791001--dc23
LC record available at https://lccn.loc.gov/2020054006
LC ebook record available at https://lccn.loc.gov/2020054007

ISBN: 978-0-367-23274-0 (hbk)
ISBN: 978-1-032-00621-5 (pbk)
ISBN: 978-0-429-27908-9 (ebk)

Typeset in Times New Roman
SPi Global, India

Contents

Figures

Tables

List of boxed case studies

Preface and acknowledgements

Place attachment, identity, and sense of place are a personal and important thing whether it be for those who live in a place or those who visit. The present volume is written by three people who, in terms of mobility at least, have 'been around'. Chris was born in China, studied in Australia, visited the United States and Europe, and has lived and worked in the UK and New Zealand; Michael was born in the UK, migrated to Australia, studied in Canada, and now works in New Zealand, Sweden, and Finland; Girish was born in Mauritius, studied in South Africa, the United States and New Zealand, worked in France, and now works in New Zealand. All three of us travel to maintain our relationship to people and place, while also working with colleagues and friends in different parts of the world. Therefore, in a very real sense, even though we primarily work and reside in New Zealand, we all have multiple homes and place attachments, with possibly more to come.

For many people our lives are not unusual. In a globalised world migration, travel, and economic and social relations that stretch over the planet are the norm. Having partners whose family are from a different place than where you were born is also common and often complex; this also further expands place attachments and relations given that, for many people, home is where the heart is. While for those who travel for work and/or pleasure home, on any given night, may also be where you are sleeping. It is in this fluid and mobile social space which is beset by the fixity of place that this book is positioned. Tourism both affects place and is affected by it. Tourism is also an extremely important part of contemporary mobility in which the significance of the place and the interplay between tourism, place, and other mobilities is underappreciated. It is therefore into this space that we would like to take you as part of a professional and personal journey into better understanding our touristic, relational, and residential selves as well as the various places that we connect and attach too.

Chris would like to firstly thank his PhD supervisors Larry Dwyer and Tracey Firth for guiding him into the field of sense of place and place attachment, giving him valuable opportunity to pursue the truth about human–place relationships. His thanks also go to his colleagues particularly Barry Ardley, Rohail Ashraf, Gary Bosworth, Jian Chen, Giacomo Del Chiappa, Jörg Finsterwalder, Stefan Gössling, Jenny (Jiyeon) Lee, Scott McCabe, Shanshan Qi, Yong Rao,

Tina Šegota, Nick Taylor, Xueli Wang, Kangkang Yu, and Chaozhi Zhang. Chris would further express his gratitude to his co-authors for their numerous insightful conversations over coffee about research and truth of life. Lastly but not least, Chris would like to thank his parents and family for their support to his academic career over the years, and particularly his daughter Emma for giving his life more meaning and happiness.

Michael would like to thank a number of friends and colleagues with whom he has undertaken related conversations and research over the years. In particular, thanks go to Bailey Adie, Alberto Amore, Dorothee Bohn, Tim Coles, Hervé Corvellec, David Duval, Martin Gren, Stefan Gössling, Johan Hultman, Tyron Love, Dieter Müller, Jan-Henrik Nilsson, Youri Oh, Yael Ram, Anna Laura Raschke, Jarkko Saarinen, Dan Scott, Anna Dóra Sæþórsdóttir, Allan Williams, Kimberley Wood, and Maria José Zapata-Campos for their thoughts on tourism and place, as well as for the stimulation of Fiona Apple, Beirut, Paul Buchanan, Bill Callahan, Nick Cave, Bruce Cockburn, Elvis Costello, Stephen Cummings, David Bowie, Ebba Fosberg, Mark Hollis, Aimee Mann, Larkin Poe, Vinnie Reilly, Dax Riggs, Henry Rollins, Matthew Sweet, Emma Swift, TISM, Henry Wagon, and The Guardian, BBC6, JJ, and KCRW – for making the world much less confining. Special mention must also be given to the Malmö Saluhall; Balck, Packhuset, and Postgarten in Kalmar; and Nicole Aignier and the Hotel Grüner Baum in Merzhausen for the pleasures of place they provide. Finally, and most importantly, Michael would like to thank the Js and the Cs who stay at home and mind the farm.

Girish would like to thank a number of colleagues with whom he has undertaken sense of place and place attachment related research for the last 11 years. In particular, thanks go to Sameer Hosany, Giacomo Del Chiappa, and Chris Ryan. More recently, working on the topic of sense of place and disasters with the following colleagues, Lucie Ozanne and Sam Spector, has further accentuated the importance of rebuilding sense of place for communities post-disaster. Girish would like to thank family, friends, and other colleagues who have supported him over the years in this academic journey.

We also wish to gratefully acknowledge the help and support of Jody Cowper for proofreading and editing. Finally, we would like to thank Emma Travis, Lydia Kessell, and all at Routledge for their continuing support.

1 Place matters!: Introduction to sense of place and place attachment in tourism

Introduction

Place is one of the most fundamental concepts in the social sciences and is critical to how we understand tourism, which is concerned with voluntarily leaving one place for another for limited periods of time. We live in a place, we experience place, and place is part of our identity and is used to describe who and what we are. One of the first questions that people ask when they meet you is 'Where are you from?'. And, when you are travelling and you are talking to locals, this is often followed by 'Do you like it here?'.

This book is therefore about the significance of place in tourism, but more so that it is specifically interested in sense of place and place attachment. These are notions of place that are imbued with deep and personal meaning and which reflect how people come to understand, present, and perceive place and how place is promoted and communicated, especially by the tourism industry, to others. Place is different from location and space. A location is a geographical marker, which tells you where A is in relation to B and C. Locations can be identified in space. But space is more abstract than place and refers to the geographical environment which can be understood in a variety of different ways including systematic descriptions of human activities and different resources. Place is space imbued with meaning.

> Place… has more substance than the word location suggests: it is a unique entity, a 'special ensemble' … it has history and meaning. Place incarnates the experiences and aspirations of a people. Place is not only a fact to be explained in the broader frame of space, but it is also a reality to be clarified and understood from the perspective of the people who have given it meaning.
>
> (Tuan, 1979, p. 387)

Place is clearly integral to tourism and is connected to both the selection of the destination that is visited by the tourist as well as the source region that tourists have come from and is their place of permanent residence. However, tourism studies are not just about the tourist but are also concerned with the destination community and those places that tourism affects. Figure 1.1 details some of the main focal points in tourism for considering ideas of place and place attachment. Tourist-oriented research on sense of place and place attachment tends to focus on the tourist's relationships to place as well as on place marketing and branding and is a recurring theme in the following chapters. From a destination perspective resident relationship to place is important with respect to how individuals and communities respond to the changes brought about by tourism, or by other factors, such as climate change. These responses are also significant because residents become tourists, and therefore the changes in the home environment and how place is perceived also affect tourist travel behaviour. The final element is that of place making. Although placemaking should ideally be undertaken with the resident as the primary focus (discussed in more detail below), placemaking and the design and purposing of space, such as making places more walkable and attractive, clearly also have implications for tourism as they provide attractive places for visitors as well as the locals.

This chapter provides a brief introduction to the book. It first examines definitions of sense of place and related concepts, before briefly looking at how place is made, promoted, and constructed.

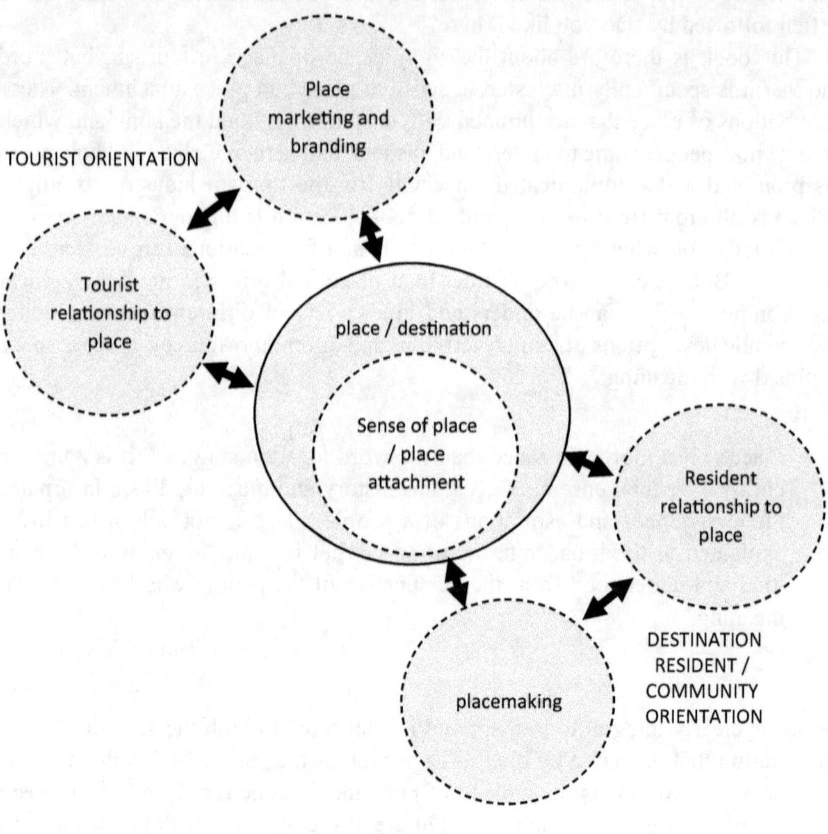

Figure 1.1 Tourism and place relationships

Definitions of sense of place

The fundamental concept that depicts the human–place relationship is that of sense of place, which is related to ideas of place attachment, topophilia, insidedness, and community sentiment (Chen, Dwyer, & Firth, 2014a, 2014b). Table 1.1 details some of the definitions of sense of place and cognate concepts taken from different disciplinary perspectives. Sense of place has been defined in multiple disciplines, including anthropology, environmental psychology, geography, landscape studies, and sociology (Cross, 2015). Although it is something of a gross generalisation, sense of place and topophilia were originally more examined through a qualitative lens, especially in the work of Tuan (1972, 1974, 1977, 1980, 1991) and its roots in humanistic geography, while place attachment has often been examined more quantitatively (Williams & Vaske, 2003). However, over time, the concepts, as with the means of analysis, have become relatively interchangeable.

One of the most important concepts in humanistic geography is that of a 'sense of place'. A sense of place arises where people feel a particular attachment or

Table 1.1 Definitions of sense of place-related concepts in different disciplines

Terminology	*Discipline*	*Definition*	*References*
Place attachment	Anthropology	Symbolic relationship formed by people giving culturally shared emotional/affective meanings to a particular space of piece of land that provides the basis for the individual's and group's understanding of, relation to, and bonding with the environment.	Low and Altman (1992, p. 165)
Place attachment	Environmental Psychology/ Geography	A complex, multivalent, dynamic experience based on the complex processes of place.	Seamon, 2014 cited in Cross (2015, p. 494)
Topophilia	Geography	Affective bond between people and place or setting.	Tuan (1974, p. 4)
Sense of place	History/ Landscape Studies	Something that we ourselves create in the course of time, the result of habit or custom.	Jackson (1994, p. 151)
Sense of place	Sociology	People's subjective perceptions of their environments and their more or less conscious feelings about those environments.	Hummon (1992, p. 254)

personal relationship to an area in which local knowledge and human contacts are meaningfully maintained. 'People demonstrate their sense of place when they apply their moral or aesthetic discernment to sites and locations' (Tuan, 1974, p. 235). However, people may only consciously notice the unique qualities of their place when they are away from it or when it is being rapidly altered, for example, as the result of rapid development or the impacts of a disaster.

The conservation of heritage is often a reaction to the rate of physical and social change within a community, and elements of the townscape or environment, such as buildings or landscape, may be conserved in order to retain elements of community identity and public memory. Generally, when people feel they are in control of their own destiny they have little call for nostalgia. However, the strength of environment and heritage conservation organisations in developed nations is perhaps a reflection of the desire to retain a sense of continuity with the past (Lowenthal, 1975, 1985). In addition, the protection of historic buildings and the establishment of heritage precincts can also effect a significant economic return to destinations because of the desire of many visitors to experience what they perceive as authentic forms of tourism.

Pacione (2001, p. 353) observed that 'certain places are regarded as distinctive or memorable through their unique characteristics or imageability and so have a strong sense of place'. This may be explained in terms of the attachments people have to a specific place, gained through experience, memory, and intention, a condition described by Tuan (1974) as topophilia. Topophilia is defined as including all of human affective ties with the environment (Tuan, 1974). Such issues are relevant for tourism as place attachment can be a significant indicator of future visitation and positive word-of-mouth. Place attachment is usually regarded as interchangeable with sense of place. Kyle et al. (2005, p. 250) define place attachment as 'the extent to which an individual values or identifies with a particular environmental setting' and it depends on 'developing clusters of positive cognitions linked to the meaning of specific places' (Jack, 2010, p. 758).

Sense of place can be conceptualised in terms of two main dimensions: (1) *relationship to place* which refers to the different ways that people are linked to places and the types of bonds (genealogical/historical, narrative/cultural, economic, ideological, cosmological, and dynamic); and (2) *place attachment* which reflects the depth and types of attachments to one place (place identity, place dependence, affective attachment, social bonding, place memory, place expectation) (Chen et al., 2014a, 2014b; Cross, 2015). This framework is illustrated in Figure 1.2.

Dimensions of place attachment, such as place identity, place dependence, and affective attachment, are also discussed in the following chapters in relation to each type of sense of place. Place dependence and place identity in particular have received substantial attention in the literature. Place dependence relates strongly to the value of place as a setting that fulfils psychological and functional needs (Smith, Siderelis, & Moore, 2010) that are not fulfilled by other locations (Stokols & Shumaker, 1981). Although related, place identity is connected more to emotional meaning (Williams & Vaske, 2003), Proshansky, Fabian, and Kaminoff (1983, p. 61) suggest that place identity is the 'combination of attitude,

SENSE OF PLACE

Relationship to places	**Place attachment**
Genealogical	*Place identity*
Narrative	*Place dependence*
Economic	*Affective attachment*
Ideological	*Social bonding*
Cosmological	*Place memory*
Dynamic	*Place expectation*

Figure 1.2 Concept structure of sense of place

values, thoughts, beliefs, meaning and behaviour tendencies, reaching far beyond emotional attachment and belonging to particular place'.

Place making, place-making or placemaking?

People are active participants in the making of places. This occurs at different scales ranging from the home to the street and then to the wider community and place of residence. Lew (2017) sees place making as an innate human behaviour. According to the Project for Public Spaces (PPS) (2007), a leading advocate for placemaking, 'for those who feel intimately connected to the places in their lives' the process of placemaking strengthens 'the connection between people and the places they share' (Box 1.1 details the 11 principles of placemaking identified by PPS). For PPS placemaking can therefore be defined as

> … a collaborative process by which we can shape our public realm in order to maximize shared value. More than just promoting better urban design, placemaking facilitates creative patterns of use, paying particular attention to the physical, cultural, and social identities that define a place and support its ongoing evolution.
>
> With community-based participation at its center, an effective placemaking process capitalizes on a local community's assets, inspiration, and potential, and it results in the creation of quality public spaces that contribute to people's health, happiness, and well being.
>
> (Project for Public Spaces, 2007)

Place making is an overarching umbrella concept that ranges from the organic and unplanned actions of individuals, which Lew (2017) defines as 'place-making,' to planned and intentional theming by governments, tourism authorities, and developers which is defined as 'placemaking' (see also Strydom, Puren,

Box 1.1 Project for Public Space principles of placemaking

1. *The Community is the Expert* – 'The important starting point in developing a concept for any public space is to identify the talents and assets within the community'.
2. *Create a Place, Not a Design* – 'The goal is to create a place that has both a strong sense of community and a comfortable image, as well as a setting and activities and uses that collectively add up to something more than the sum of its often simple parts'.
3. *Look for Partners*
4. *You Can See a Lot Just By Observing.*
5. *Have a Vision*
6. *Start with the Petunias: Lighter, Quicker,* Cheaper – 'The complexity of public spaces is such that you cannot expect to do everything right initially. The best spaces experiment with short term improvements that can be tested and refined over many years!'
7. *Triangulate* – 'Triangulation is the process by which some external stimulus provides a linkage between people and prompts strangers to talk to other strangers as if they knew each other' (Whyte, 1988, p. 94), i.e. events in public spaces that focus the attention of the users of the space, both permanent residents and visitors, on an event or object and facilitate communication between otherwise disconnected strangers (Karahalios & Donath, 2004).
8. They Always Say 'It Can't Be Done' – 'Creating good public spaces is inevitably about encountering obstacles, because no one in either the public or private sectors has the job or responsibility to "create places".
9. *Form Supports Function*
10. *Money Is Not the Issue*
11. *You Are Never Finished*

See: Project for Public Space, Eleven principles for creating great community places, https://www.pps.org/article/11steps

& Drewes, 2018, on theoretical developments in placemaking). The former Lew (2017) regards as being more local in scope while the latter he frames as being more global. Both use essentially the same tools, for example, design and intangible elements such as branding, but Lew suggests that the intentions and outcomes are vastly different.

Tourism is a socio-economic phenomenon that is a major contemporary force in (re)producing places and culture (Hultman & Hall, 2012). According to Lew

(2017, p. 448), "Tourism development, as an economic activity, almost always has a neoliberal planned placemaking agenda. Organic place-making emerges through individual agency, which if allowed to proceed, will make its imprint even in a heavily master planned tourism landscape". An example of placemaking in a tourism context is that of Barcelona (Mansilla & Milano, 2019), which included new cultural icons and tourist attractions embody so as to symbolise and communicate the city to outsiders in order to boost the city's visitor activities. Indeed, the development and promotion of cultural quarters and precincts or other special designations of areas of a city are a classic economically focused placemaking strategy.

An example of place-making is provided by Iaquinto (2020) with respect to the role of backpackers and other mobile populations whose individual actions, as opposed to industry directives, contributed to making places. He found that '[p]laces inhabited by backpackers were in constant flux and 'co-created' via practices in conjunction with an array of other phenomena. As backpackers interacted with one another, other people and the various materials, temporalities and environments that were present' (Iaquinto, 2020, p. 1). Nevertheless, more enlightened jurisdictions that focus on the immediate needs of its residents will often embrace placemaking as a way of improving the quality of life for local people, while at the same time realising that making a place more attractive for people to live will also be attractive for mobile populations such as tourists, migrants, and students. The characteristics of community-driven and bottom-up form of place making are detailed in Table 1.2.

PPS also come up with a checklist of items that they identify as comprising a great space (Table 1.3). This is composed of key attributes and the intangibles

Table 1.2 Characteristics of place making

What placemaking should be	*What place making is not*
• Community-driven	• Top-down
• Visionary	• Reactionary
• Function before form	• Design-driven
• Adaptable	• A blanket solution or quick fix
• Inclusive	• Exclusionary
• Focused on creating destinations	• Car-centric
• Context-specific	• One-size-fits-all
• Dynamic	• Static
• Trans-disciplinary	• Discipline-driven
• Transformative	• One-dimensional
• Flexible	• Dependent on regulatory controls
• Collaborative	• A cost/benefit analysis
• Sociable	• Project-focused

Source: Project for Public Spaces (2007).

Table 1.3 Project for Public Spaces: What makes a great place?

Attributes	*Intangibles*	*Measurement*	*Questions*
Access & Linkages The accessibility of a place is judged by its connections to its surroundings, both visual and physical. A successful public space is easy to get to and through; it is visible both from a distance and up close. The edges of a space are also important.	Continuity Proximity Connected Readable Walkable Convenient Accessible	Traffic data Mode splits Public transport/ Transit usage Pedestrian activity Parking usage patterns	Can you see the space from a distance? Is its interior visible from the outside? Is there a good connection between the space and the adjacent buildings, or is it surrounded by blank walls? Do occupants of adjacent buildings use the space? Can people easily walk to the place? For example, do they have to dart between moving cars to get to the place? Do sidewalks lead to and from the adjacent areas? Does the space function for people with special needs? Do the roads and paths through the space take people where they actually want to go? Can people use a variety of transportation options – bus train, car, bicycle – to reach the place? Are public transport/transit stops conveniently located next to destinations such as libraries, post offices, park entrances, etc.?
Comfort & Image Comfort includes perceptions about safety, cleanliness, and the availability of places to sit	Safe Clean Green Walkable Sittable Spiritual Charming Attractive Historic	Crime statistics Sanitation rating Building conditions Environmental data	Does the place make a good first impression? Are there more women than men? Are there enough places to sit? Are seats conveniently located? Do people have a choice of place to sit, either in the sun or shade? Are spaces clean and free of litter? Who is responsible for maintenance? What do they do? When? Does the area feel safe? Is there a security presence? If so, what do these people do? When are they on duty? Are people taking pictures? Are there many photo opportunities available? Do vehicles dominate pedestrian use of the space or prevent them from easily getting to the space?

Attributes	Intangibles	Measurement	Questions
Uses & Activities Activities are the basic building blocks of great places. They are the reasons why people visit in the first place, and why they return. They are also what makes a place special or unique. When there is nothing to do in a place, it will sit empty and unused, a sign that something needs to change.	Fun Active Vital Special Real Useful Indigenous Celebratory Sustainable	Local business ownership Land use patterns Property values Rent levels Retail sales	Are people using the space or is it empty? Is it used by people of different ages? Are people in groups? How many different types of activities are occurring – people walking, eating, playing baseball, chess, relaxing, reading? Which parts of the space are used and which are not? Are there choices of things to do? Is there a management presence, or can you identify anyone is in charge of the space?
Sociability This is a difficult quality for a place to achieve, but once attained it becomes an unmistakable feature. When people see friends, meet and greet their neighbours, and feel comfortable interacting with strangers, they tend to feel a stronger sense of place or attachment to their community – and to the place that fosters these types of social activities.	Diverse Stewardship Cooperative Neighbourly Pride Friendly Interactive Welcoming	Number of women, children and elderly Social network Volunteerism Evening use Street life	Is this a place where you would choose to meet your friends? Are others meeting friends here or running into them? Are people in groups? Are they talking with one another? Do people seem to know each other by face or by name? Do people bring their friends and relatives to see the place or do they point to one of its features with pride? Are people smiling? Do people make eye contact with each other? Do people use the place regularly and by choice? Is there a mix of ages and ethnic groups that generally reflect the community at large? Do people tend to pick up litter when they see it?

Source: Project for Public Spaces, 1975–2018, 2007.

that come from place and which, in turn, contribute to sense of place and the development of place attachment. From a planning perspective they also identify a number of metrics that can be used to measure place attributes. As noted above these can be used for both top-down and bottom-up place making and are relevant to permanent and temporary residents as well as tourists. However, for non-residents, such as tourists, to become aware of the attributes of place it must be communicated to them. The significance of place marketing is noted in the next section.

Place marketing

Place marketing is integral not only to how places attract tourists but also to how they may attract and retain other desired mobile elements, such as capital, firms, and people in an effort to encourage regional economic development (Kavaratzis & Ashworth, 2008). Place branding and marketing are some of the main means by which the sometimes latent demands and interests of consumers are converted into place awareness and, potentially, into purchasing a place product, whether it be experiencing a particular destination, a tourist or a migrant, or purchasing food or a good that is specifically associated with a place (Hall, 2014a). Tourism is obviously deeply connected to the place marketing process not only because of the efforts of destination and tourism marketing to attract tourists but also because of how it is used as a focus for urban revitalisation and redevelopment processes (placemaking), and promotion strategies that put a strong emphasis on the deliberative social construction of place for consumptive purposes. However, it should also be noted that place marketing will also take advantage of organic elements (place-making) of place image as part of the elements of formal campaigns, although there are dangers of such elements becoming commoditised in the longer term.

Although, as Berglund and Olsson (2010) observe, it is inappropriate from a marketing perspective to equate place marketing with selling, place marketing is often used interchangeably in the social sciences with the concepts of place branding, place selling, and place imaging and reimaging (Berglund & Olsson, 2010). Place marketing is therefore somewhat of an umbrella term to describe a wide range of formal, usually commercially and economically driven, place-related communications. Importantly, with respect to the relationship between place marketing and placemaking, both are fundamental to place competitive strategies, i.e. when one place is deliberately trying to outcompete other places in terms of tourists, investments, and/or exports (Hall & Williams, 2020). Place branding is the development and promotion of a spatially framed brand in order to differentiate a place from other locations so as to gain brand value advantage for its firms, organisations, people, products, and services (Hall, 2014a).

Place marketing and branding are undertaken via a mix of material and intangible practices (Morgan, Pritchard, & Pride, 2011). Material strategies include mechanisms such as placemaking, the development of flagship projects, and the hosting of hallmark events. Immaterial strategies include the use of advertising, branding, slogans, media placement, public relations, and the development of new place myths and stories that can then be reproduced for commercial and economic objectives (Hall, 2014a). The notion of constructed advantage through the integration of public and private economic interests is therefore critical in identifying the key role that place branding and marketing play as competitive practices in a tourism context. Significantly, sense of place is an integral component of both the immaterial and material strategies of place marketing and branding and is also directly affected by them, as well as other factors such as environmental change.

Box 1.2 Santa Claus, place branding and sense of place: Will climate change kill Santa Claus?

Santa Claus (also referred to as Saint Nicolas and Father Christmas) has been described as the world's strongest brand (Arruda, 2003): 'Santa Claus is the envy of brand managers everywhere. His brand attributes are clear and desirable to virtually everyone. Even parts of the world that have no connection to the holiday know who he is and what he stands for'. Along with polar bears and reindeer, Santa Claus is also a significant symbol of the Arctic and the 'North Pole'.

Although there is much attention given to Santa as a part of contemporary consumption practices, even in non-Christian and secular contexts, Santa is also important as a place brand and has become integral to the competition between a number of locations to be the home of Christmas. Prestes (1995), for example, discussed the development of "the Santa Claus industry" as a form of post-modern tourism with respect to the Santa Claus Village in Rovaniemi, Finland. Although he noted that an essential element of the Rovaniemi strategy was 'convincing the rest of the world that Finland, and Lapland in particular, was the real home of Santa Claus–against rival claimants in Alaska, Sweden, Norway, and Greenland'.

Table 1.4 provides examples of the way that Santa Claus, Father Christmas, and related myths are used for place branding purposes (Hall, 2008a, 2014a). The table is not exhaustive and represents more overt examples of the use of Santa Claus in high-latitude place branding and competition. The table therefore does not include theme parks such as Santa's Workshop, North Pole, Colorado; Santa Claus, Indiana; nor the villages of Demre and Patara in Anatolia, Turkey, which were homes of the Bishop of Myra – the historical source of St. Nicholas and the myth of Santa Claus. Similarly, it does not include the promotion of Christmas tourism in Bethlehem, Palestine, which focuses more on non-Santa, Christian, dimensions of Christmas in tourism marketing and development.

Finnish Lapland has been described as the 'Santa tourism superpower' (Snellman, 2006) because of the extensive promotion of Santa Claus as a regional and national brand. Focused initially on Rovaniemi, the self-proclaimed 'Christmas capital of the World' and 'hometown of Santa

Table 1.4 Examples of Santa Claus related place branding

Country	*Location*	*Christmas character/brand*	*Material evidence of brand*
Canada	North Pole H0H 0H0 (official Canadian postal code for mail to Santa Claus)	Santa Claus (In English); Father Christmas (In French)	Canada Post receives more than a million letters and emails for Santa some years. These are handled by volunteers.
Finland	Rovaneimi	Santa Claus; 'Rovaneimi –the Christmas City', 'Christmas capital' 'Rovaniemi, Santa's official hometown'	Santa Claus' Main Post Office received over half a million visitors and many letters each year; also home to Santa Park; Santa Claus Village and other Christmas-related attractions, such as reindeer burger.
Finland	Saariselkä	Santa's Resort	Santa's Resort
Finland	Turku	'Christmas city of Finland'	Christmas Street, festival events
Greenland	The North Pole (Greenland as gateway)	Santa Claus	Nuuk has a Santa's post office and post box.
Iceland	Dimmuborgir in Myvatnssveit	Santa Claus, Yule Lads	Christmas market, Santa Claus post office, Santa in residence from late November
Norway	Drøbak	Julenissen (Christmas elf) also nisser	Julehuset (the Christmas House) (10 months a year), post office
Norway	Savalen	Julenissen (Christmas elf) also nisser (gnomes)	Santa Street, Santa's house and workshop (seasonal); Santa Association and Winter Games
Sweden	Mora (Dalarna)	Santa Claus	Santa World – Tomteland opened in 1984
USA	North Pole, Alaska (near Fairbanks)	"Where the Spirit of Christmas lives Year round"	Santa Claus House, 12 m Santa statue, Santaland RV Park, Santa's Suds Laundromat, street names, Post Office

Source: After Hall (2008a, 2014a).

Claus' which has a great concentration of Santa- and Christmas-related attractions (Tervo-Kankare et al., 2013), Santa tourism has also become an important Christmas season tourism activity for a number of other locations in Lapland, while Finland has also promoted itself as 'the Christmas Country' at various times and Finnair has also promoted itself as the official airline of Santa Claus since 1983.

Santa Claus tourism is of great importance to the Lapland economy, and Santa Claus tourism demonstrates the powerful influence that the development of new place myths and stories can have on tourism development and marketing, as well as the tourists who visit. However, Santa tourism in the region is facing two substantial challenges with respect to the Santa and place relations. First, within Lapland there is increasing internal competition between locations associated with Santa. Although Rovaniemi is the market leader, the rise of international travel to other Finnish destinations to meet Santa presents substantial challenges to tourism in Rovaniemi and its role as a gateway. Travel operators to other Lapland destinations increasingly promote holiday packages as being more 'authentic' than that available in the tourist hub of Rovaniemi. In the longer term the fact that there are so many Santa tourism options available in Lapland may potentially even call to question the uniqueness of the Lapland Santa experience unless it is well managed.

The issue of commodification and perceived authenticity of the Santa experience is obviously one dimension that is already starting to influence travel companies in their destination choice (Hall, 2008a; Tervo-Kankare et al., 2013). However, more importantly in the longer term will be assurances to the customer that some of the environmental images that underlie the brand images and values of Christmas in northern high latitudes, such as snow, reindeer, and conifers (Grenier, 2007; Tervo-Kankare et al., 2013), will be available for visual and experiential consumption.

In northern Finland, as in other winter tourism locations, especially in the Arctic and the Nordic countries (Demiroglu & Hall, 2020; Hall & Saarinen, 2020), increased concern is being expressed about the impacts of global heating and the climate crisis, including the effects on 'snow security'. The Christmas tourist season in Rovaniemi is increasingly portrayed as a potentially riskier destination than other resort areas in northern Finland (Falk & Vieru, 2019; Hall, 2008a; Tervo-Kankare et al., 2013). Indeed, the snow-deficient and somewhat warm winters in the mid-2000s have already 'led travel organizers and tourism entrepreneurs to develop destinations further north in case Rovaniemi region does not have snow early enough' (Tervo-Kankare et al., 2013, p. 312). In Finland, the annual mean temperature is now about 2.3° C higher than it was in mid-nineteenth century. Depending on the scenario adopted, the annual mean temperature is projected to increase by 1.9–5.6° C by the 2080s, compared to the period of 1981–2010. At the same time, the mean annual precipitation is forecast to increase by 6%–18% for the same period. Significantly, for winter and Santa Clause tourism, the changes are expected to be larger during the winter than the summer months (Venäläinen et al., 2020). Importantly, such climate change and associated environmental change will affect both residents' and visitors' sense of place.

Research has already indicated the adaptation of winter tourism businesses to climate change in Finland. Nevertheless, a survey of Christmas tourists to Rovaniemi found that less than one-fourth of the respondents considered Rovaniemi as an appealing destination if it had no snow, while a slight majority indicated that they would not be willing to travel to Rovaniemi in January if the snow season would not start until then (Tervo-Kankare et al., 2013). The weather-dependent elements of a winter Christmas, such as snow, cold, and snow on trees, were critical to respondent's attitudes as to intentions (Tervo-Kankare et al., 2013). The long-term implications of climate change for northern place branding and the imagined landscapes of Christmas are significant, especially as the November–December period is more vulnerable to the effects of climate change than January–February. Substantial emphasis has been placed on promoting the Christmas brand, yet the potential effects of not having snow at Christmas time is not just one of providing substitute activities or using artificial snow for trails (Tervo, 2008); it is also one of the fulfilling place brand image and values. Lapland and Finland are promoted with images of snow and reindeer for the Christmas season and natural snow, as opposed to rain or artificial snow, is integral to many a child's (and adult's) understanding of what constitutes a 'real' Christmas.

The 'Christmas card' images of Santa also incorporate other elements of high-latitude imagery such as pine trees (Carpino et al., 2018; Venäläinen et al., 2020) and reindeer (Vors & Boyce, 2009) that are also under substantial threat and undergoing decline as a result of either the direct or indirect effects of climate change. Although the branding of destinations with respect to Christmas and Santa Claus is socially constructed, it is underlain by elements of the physical environment that have become part of the imaging of a place as the 'home of Santa' via both place marketing and via the promotion of Christmas overall. For example, approximately half of the Christmas card markets in countries such as Australia and the USA feature winter scenes (Hall, 2014a). Such social imaginaries are therefore substantial contributions to the sense of place felt by visitors to northern Finland, as well as other northern latitude locations in which Santa is central to place image and marketing.

Although elements of the Christmas place myth with respect to Santa's workshop or house, and characters such as Mrs Claus or Santa's helpers can be reconstructed in almost any location, key environmental elements such as reindeer, pine trees, and the white wintered landscape of a high-latitude northern hemisphere Christmas are much harder elements to reproduce in an authentic fashion. Undoubtedly, destinations can also promote Summer visitation. However, even success at that time of year depends on the believability of the location as a home of Santa from its winter attributes. The notion of Santa living in a wintered white landscape of snow at Christmas time which is represented in advertising and millions of seasonal cards and messages each year cannot be easily overturned. Therefore, the environmental attributes that are integral to the Santa myth in consumer society and which underlie Christmas place marketing may clearly be affected by climate change to the likely detriment of destinations that currently rely on Santa place brands and their associated visitor place attachment and sense of place.

Source: Adapted from Hall (2014a).

Conclusions

In recent years, emphasis on virtual worlds and cyber-space and the promise of e-tourism has led to a downplay of the importance of place. However, nothing could be further from the truth. Instead, place is arguably more important than ever, especially in terms of tourism. Conventional notions of place, particularly our sense of place, transcend geographical barriers of distance and advances in media and information technology have created globally aware consumers for whom place matters and for whom there is now a much broader set of destinations that they may wish to visit. As a result place attributes and how they are communicated and understood have become integral to tourism as the industry has sought to differentiate destinations on the basis of their place characteristics – both real and imagined.

Sense of place and place attachment are also of significance to tourism development for several reasons. The redevelopment and reimaging of places and communities for tourism purposes may force long-term residents to leave as a result of new economic conditions and will often change the nature of the destination community as place is restructured. In these instances, the identification of residents with the physical and social structure of the neighbourhood may be deeply disturbed, leading to a condition of what Relph (1976) aptly described as 'placelessness'. Residents of destinations who find themselves faced with rapid tourism development may therefore attempt to preserve components of physical and social infrastructure in order to retain their sense of place. However, this does not mean that tourism is necessarily damaging to communities. Tourism can offer new economic opportunities and, if planned appropriately, can even reinforce existing feelings of place attachment. Yet to do so requires greater community control over tourism and a bottom-up approach to place making and a great acknowledgement of the relationship between tourism and place. As Britton (1991, p. 475) recognised more than 30 years ago,

> An analysis of how the tourism production system markets and packages people is a lesson in the political economy of the social construction of 'reality' and social construction of place, whether from the point of view of visitors and host communities, tourism capital (and the 'culture industry') or the state – with it diverse involvement in the system.

This book therefore aims to fill some of the substantial gaps (Dupre, 2019) that exist with respect to understanding the relationships between place and tourism. As noted in the introduction place is a concept that is rich in social and other relations and represents the personal attachment that we all feel with where we live and visit, and the places we would like to visit. The series of chapters go through some of the major place concepts as well as new developments of place understanding in light of virtual worlds, spirituality, and disaster. As numerous works have suggested 'place matters' (Alkon & Traugot, 2008; Beech, 2014; Garfinkel, 2019; Jack, 2010; Malatzky et al., 2020; Pares, Marti-Costa, & Blanco, 2014; Rickly-Boyd, 2013; Strom, 2017; Thompson, 2017; Wasfi, Steinmetz-Wood, & Kestens, 2017), hopefully this work further emphasises this point in a tourism context and, in terms of developing a better understanding of place in a mobile and connected world, how tourism is significant for place.

2 Where I am from

The genealogical sense of place

Introduction

No matter how sense of place is defined, it reflects the nature of connecting an individual to a place, whether in the real or virtual world, and this connection can be related to one's *insideness* mentally and emotionally. Places and physical environments imbued with symbolic meanings are essential in defining the self. Two of the most important symbolic meanings are origin and sense of belonging.

Place identity, following this logic, can be defined as a sub-structure of personal identity of an individual that is based on cognition of the combination of the physical and social settings in which the individual lives. From this sense, place identity is determined in the creation of sense of place, by the place, and its characteristics, and provides contentment in biological, genealogical, historical, psychological, social, and cultural senses (Proshansky, Fabian, & Kaminoff, 1983). In many cultures, such as that of China and the Māori of New Zealand, one's origin is considered essential in making sense of the individual–place relationship, via an affective or evaluative process. However, this is something common to many cultures as one of the questions you often ask people when you first meet them and you try and 'place them' is 'Where are you from?'.

This chapter examines several related concepts such as place identity, sense of place, and place attachment in relation to how they contribute to feelings of rootedness and belonging and thus shape an individual or community relationship with place. We examine these concepts and relationships from the lens of sense of place emanating from genealogical relationships.

The chapter starts with a discussion of the concept of rootedness and then examines what rootedness means and its relationships with sense of place, sense of belonging, and sense of ownership. The chapter then examines rootedness from the perspective of Chinese culture and evaluates how Chinese culture has been represented for tourism purposes and what this means for sense of place. Thereafter, the chapter progresses to examine how cultural identity of indigenous populations such as the Māori has been appropriated and represented for tourism purposes and the implications for sense of place. By focusing on the genealogical aspects of sense of place, we highlight similarities and differences in sense of place and place attachment for individuals from different cultures such as Chinese and Māori. The chapter concludes by comparing Chinese and

Māori cultures to illustrate how rootedness develops and subsequently informs meaning creation from a cultural perspective and how this meaning is marketed for tourism purposes.

Rootedness, sense of place, sense of belonging, and sense of ownership

How one understands place is often benchmarked to children's relationship with their parents and caregivers, and especially their mothers. If place is defined broadly as a focus of value, of nurture, and support, the mother can be considered as children's primary place (Tuan, 1977). Similar to the mother to a child, the first place one lives in may well be the first enduring and independent object in this individual's world of views. The first place where an individual lives is often 'unique', for it plays a distinguished role in forming the first ever perceived experience with a place. In this experience creation, one becomes attached to objects other than significant persons, and to localities, and the idea of place becomes more specific and geographical along with this process (Tuan, 1977).

In a child's case, the geographical horizon expands as the child grows, and a sense of place is gradually built, since this place can acquire deep meanings through the steady accretion of sentiment over the years of growth. Living experiences, imaginative stories, told stories, and the general process of making sense of the world contribute in the creation of sense of the place. This place, covering the important concept of 'home', is regarded by Tuan (1977) as being at the cosmic centre of an astronomically determined spatial system from which the idea of rootedness is developed.

Rootedness in a place implies continuity in place, reflecting a strong sense of past and continuity in place. Physically, rootedness develops from long habitation at one place and reflects one's geological connection to a place via this long habitation in generations or by ancestors. Tuan (1980) defines place rootedness as a strong and focused bond in its essence meaning being completely at home, unreflectively secure, and comfortable in a particular location. Rootedness indicates a profound attachment to homeland, which is a worldwide phenomenon not limited to any particular economy or culture. Place-based ideology, therefore, links people to place through a rooted sense of place (Hay, 1998a).

People who have stayed in a place for a short period of time may also develop a sense of place, but sense of place is more commonly fostered in long-term interactions and engagement, which requires long-term residence. Nevertheless, it is important to note that given the advent of virtual and digital communicative technologies, it may become easier to maintain long-term interactions with a place without actually physically being there all the time. Tuan (1980) proposes the concept of rootedness in describing a certain psychological state of being at home in an unself-conscious way. This specifies people who identify themselves as insiders of a place through birth and/or long residence, whose sense of place usually derives from a heightened historical consciousness. Hay (1998a) refers to this as a rooted sense of place. Rootedness often results in strong place attachment and sense of place, which is considered in a developmental continuum and relies on a relevant long term of residence (Hay, 1998b; Tuan, 1980). Rootedness is also a source driving topophilia, sense of belonging, nostalgia, and sense of ownership.

Sense of belonging

A sense of belonging is different from topohilia, which includes all human affective ties with the environment (Tuan, 1974), and instead indicates a proactive establishing and maintaining relatedness to place (Hagerty, Lynch-Sauer, Patusky, Bouwsema, & Collier, 1992). A need for belonging has been identified as a basic human need, and ranked third in Maslow's hierarchy of needs (McLeod, 2020). Anant (1966) defines belongingness as the sense of individual involvement in a socially constructed system in order for the individual to feel that they are an indispensable and integral part of the whole social system, e.g. this could be a community, including an online community, or a culture or sub-culture. In a place context, where it covers both a social system and a geographical and physical space, a sense of belonging indicates the recognition and urge of being part of this place. Hagerty et al. (1992) suggest this sense is both affective and evaluative. In a child's case, a child is born in a place, expanding his/her cosmos view upon the surroundings, naturally developing an affective sense of belonging, often subconsciously. In a transnational migration context, an immigrant evaluates a place in terms of its available economic, social, and physical resources, which all contribute to the development of a sense of belonging to this place over time (Raffaetà & Duff, 2013). For many, in general, a sense of belonging to a place is developed from both types of processes.

In the migration context, genealogy has been applied as an appealing tool for navigating through questions of belonging (Nash, 2002). This is largely because immigrants or people with high mobility often possess multiple place identifications and multiple belongings to the place(s) of residence and other places (such as place of origin if any). Specifically, an individual's genealogical origin in a place, such as family rootedness, birthplace, parenthood, a blood tie, family trees, establishes a genealogical connection between an individual and a place, where an affective and evaluative sense of belonging may be developed. This individual further develops his/her sense of place from this sense of belonging, leading to a largely genealogical sense of place (as illustrated in Figure 2.1). This sense of belonging functions as a pull factor in the human–place relationship. In a tourism context such a genealogical sense of place can encourage visits to places with the ancestors of the tourist often being many generations back. Some destinations which have been associated with large-scale migration, such as Scotland following the clearances and Ireland after the potato famine, have used genealogical or 'roots' tourism as a major factor in their destination promotion, while it also accounts for a significant promotion of visitors to some destinations (Cheer & Reeves, 2013; Higginbotham, 2012; Marschall, 2015; Pelliccia, 2018; Skipper & Davidson, 2018).

Sense of belonging is also an interactive concept, implying not only one's recognition and acceptance of a place and its social system, but also recognition and acceptance of a member by others within this place (Anant, 1966). This implies a dynamic nature of sense of belonging to a place, capturing both the changes with time, in one's making sense of the place, as well as in the place making sense of an individual. One could lose his/her sense of belonging to a place, either due to his/her own evaluation of the place, or because a place changes to something which has an unrecognisable social system.

Sense of ownership

In the interactions between an individual and a place (as well as its social system), a further hierarchy can be established in terms of the amount of resources owned and cultural and historical entitlements to the place, which derives a sense of ownership by some. One's sense of ownership, of course, often derives from residence status, land or property ownership, and legal and cultural rights to access a place. However, a psychological sense of ownership of a place may develop from genealogical origins and rootedness, and be constructed socially (Aitken & Campelo, 2011). For instance, regular repeat visitation to a place and its activities can promote attachment and familiarity and a sense of ownership (Shamsuddin & Ujang, 2008), meaning that tourists and second home owners can develop strong attachments and a sense of ownership of a place even if it is not their permanent residence (Adie, 2020; Aronsson, 2004; Nouza, Ólafsdóttir, & Sæþórsdóttir, 2018; Stedman, 2006).

In seeking to understand the emotional bonds involved in a sense of place, Fuhrer, Kaiser, and Hartig (1993) examined three processes of place attachment that position place as a facilitator of emotional needs. These three processes are based on the three different emotions of security, autonomy, and arousal. Autonomy, out of the three processes, outlines one's sense of ownership, the ability to regulate interactions with others, a sense of control, and the ability to initiate change (Inalhan & Finch, 2004). A sense of ownership can therefore facilitate the construction of a social structure that differentiates insiders (or locals or hosts) from outsiders (or guests or visitors). In such a social structure, insiders take a stronger sense of responsibility and further a sense of citizenship, illustrating power and empowerment in order to lead interactions with outsiders (Lachapelle, 2008). In consequence, these people draw the place into their bubble of responsibilities, enabling the development of a sense of control and the ability to initiate changes (see Figure 2.1). Relatively speaking, a sense of belonging may illustrate a rather passive identification and integration of oneself to a place and its system, while sense of ownership indicates often a proactive process of managing a place for one's own benefits.

Building a sense of belonging to and sense of ownership of place reflects direct and indirect interactions with a particular place and further derives a set of collectively held place image. Senses of belonging and ownership are regulated by the three basic facets of a place: (1) their functions, norms, and collectively defined identities and social roles of members, (2) personal and collective purposes

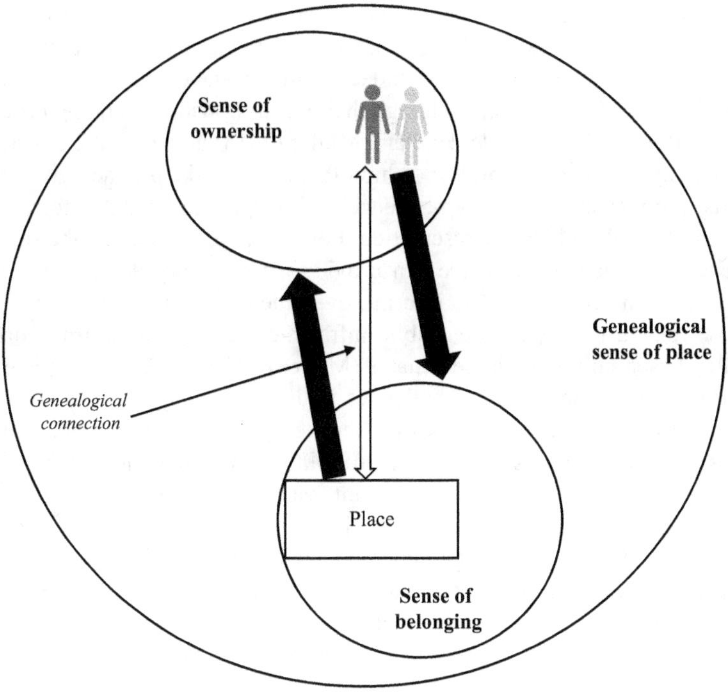

Figure 2.1 Genealogical sense of place

important to the inhabitants and often largely influenced by values handed down from ancestors, and (3) evaluations of occupants, physical features, and/or social functions (Stokols & Shumaker, 1981).

Accordingly, a sense of ownership finds its role in community development (Lachapelle, 2008), especially in regional tourism development. In many jurisdictions, local residents of a tourism destination can have significant powers over various aspects of tourism development as a result of planning processes, ranging from their general attitude and level of support (Teye et al., 2002), to their direct and indirect involvement of tourism service delivery (Chen et al., 2015). Locals often form civil societies to illustrate their sense of ownership, in supporting or opposing urban development related to tourism activities, such as the case of anti-tourism industry protest in Barcelona (Hughes, 2018).

In the summer of 2017, a large number of youths took to the streets in Barcelona, Spain, as part of the anti-tourism industry protests. At somewhere close to the famous Camp Nou stadium, a group of protesters stopped a sightseeing tour bus, slashed its tyres, and daubed '*tourism kills neighbourhoods*' on one side of the bus. These actions were taken on camera, and a video was uploaded to YouTube. In the caption, it says: '*mass tourism kills the neighbourhoods, destroys the territory and condemns the working class to misery*' (Burgen, 2017). The tension between locals and tourists has increased dramatically in the context of overtourism, in fighting over the ownership of a place (Seraphin et al., 2020). However, fights over ownership occur between locals and destination stakeholders as well, driving a careful rethinking of the sense of ownership and views regarding its 'ideal' states (Gössling et al., 2020). Indeed, different stakeholders may all have a sense of ownership of a place, although their standing within planning processes may be different. For example, Hall and Müller (2018) note how the governance and planning of second home ownership often relate to the primacy given to the declared permanent or primary residence with respect to its status for tax purposes and formal political representation. For example, in Sweden, the selection of place of permanent residence can provide for monetary flows from central to local government as well as inform government investment decisions, such as those related to health care, public infrastructure, and public transport that are often based on population estimates (Müller & Hall, 2003). Multiple dwelling also provides fundamental challenges to understandings of democratic representation because of the relationship between political standing and property ownership and rights (Ostrow, 2002). In political systems where voting is not based on property, there remain significant issues in enabling meaningful political representation for 'non-permanent' dwellers such as second home owners at the local scale, as even though they may have a sense of place ownership and even own property they may not be able to vote (Rinne et al., 2014). However, having a rootedness can also have other implications, as illustrated in the following example from Chinese culture in relation to travel and a sense of place.

Rooted sense of place

The value of a sense of belonging and a rooted sense of place exists in many cultures and regions. In Chinese culture, there is a saying '落叶归根', translated as '*fallen leaves return to the roots – to revert to one's origin*'. People are supposed to return to their homeland, to complete the circle of life.

Based on Confucian doctrines, traditional Chinese culture values ancestors and clans along with their rootedness in places, which play leading roles in political and economic affairs in ancient and modern China (Watson, 1982). Clan halls still exist in many Chinese families, enshrined and worshipped, where genealogy books are kept and maintained. From this sense of rootedness, philosophical and ethical codes have been developed and followed for generations (Sheng, 2005), serving to regulate individual and social identities in Chinese culture. Chinese clan culture (氏族文化) incorporates two parts: *zong* (宗) and *zu* (族). *Zong* refers to the common ancestors respected by their descendants, while *zu* means the collective of people who live together (Tang, 2017). Each clan in China is an independent social system that is ultimately linked to a place and largely based on kinship. Within the clan, common values, beliefs, and social norms are fostered, usually from their sense of place, and generate great implications in other associated cultures such as business culture (Chuang et al., 2012).

Extending from this clan culture, social relationships in ancient China were traditionally founded on three bases: (1) genealogy, (2) rootedness in education, and (3) rootedness in place, all of which can be traced to one's place of origin.

For instance, the Donglin, a political party with great impact on the political regime in Ming Dynasty, originated from the Donglin Academy in the Yangzi delta in Changzhou prefecture, Wuxi County, reflecting a powerful regional ideology greatly influencing imperial policy of China in history (Elman, 1989). In contemporary China, in spite of increasing mobility due to industrialisation and urbanisation, the traditional clan culture is making a comeback, driving people to trace and re-examine their kinship and derived identities. For instance, in tourism there is a growing trend of people travelling to their places of origin, so as to retain or establish virtual and physical connections to their roots and clans, leading to a significant diaspora tourism market (Coles & Timothy, 2004; Huang et al., 2016). A diaspora is

> groups of people scattered across the world but drawn together as a community by their actual (and in some cases perceived or imagined) common bonds of ethnicity, culture, religion, national identity and, sometimes, race.
>
> (Coles & Timothy, 2004, p. 3)

Individuals and groups of people have migrated from one location to another throughout history, either by force or for a number of reasons, which was summarised by Cohen (1997) into a fivefold classification including victim diasporas, labour diasporas, imperial diasporas, trade diasporas, and cultural diasporas (Cohen, 1997). Irregardless of which type of diaspora, most migrants retain strong emotional bonds to their places of origin, which acts as a strong motive to return to these places (Hall & Williams, 2002; Timothy, 2008; Williams & Hall, 2000).

Diasporas gradually develop their sense of belonging to the host place, and further their sense of ownership, via their perceptual prominence of environments, specifically including the functional, motivational, and evaluative significances of place (Stokols & Shumaker, 1981). On the other hand, these diasporic groups often retain a certain level of emotional engagement and allegiant connection to their places of origin. Behind this proactive and continuous retention of these connections is a response to these migrants' placelessness in relation to their places of origin (Farrar, 2011). A typical response to this placelessness is often a form of nostalgia. Nostalgia may be defined as 'literally a painful (i.e., physical) longing (algos) for home (nostos)' (Farrar, 2011, p. 727). It is widely acknowledged that nostalgia is a key motive for many behaviours related to place or origins, such as any intention to retain connections, revisiting homelands by migrants and their descendants, and even purchase of second homes or retirement homes. For instance, roots tourism, focusing on places with direct relation to one's own lineage or familial legacy, has long been a popular tourism segment in China and general Asian regions. Within roots tourism, genealogy-based travel illustrates tourism centring personal heritage and provides a way to explore one side of personal heritage not understood or felt at home (Timothy, 2008).

Box 2.1 The cultural significance of 落叶归根 (fallen leaves returning to the roots) and its implications in diaspora tourism

In 2007, a comedy movie *Getting Home* was released in cinema in China. This movie, in which many Chinese famous actors and iconic comedians starred, was about a story of a rural migrant worker travelling back home from a city, carrying his friend's remains to hometown to fulfil his promise. The movie is humourous and touching at the same time, putting an absurd idea of carrying a corpse through various modes of public and private transportation, which is essentially against the law in China as in most countries! The protagonist awkwardly disguised his dead friend as a drunk guy and ran into many unexpected adventures where he had to face challenges or was helped by strangers.

Behind this interesting story depicted by the movie is the significant cultural value of China that one should return to his/her root (hometown) in the end of the life journey. From a social sense, return migration is driven by the dominating social connections in China's long history of collectivism. From a psychological point of view, nostalgia is another key reason motivating people to return to their hometowns. However, the most significant driver of return migration lies in culture and cultural norms. In traditional Chinese philosophy, one must not forget his/her origin, and it is highly appraised if returning home with glories (Wang & Fan, 2006). In almost every dynasty of China, government officials are permitted and encouraged to go back to their hometowns after retirement, after serving the country in the capital or somewhere else as expatriates. In contemporary China society, travelling with the theme of searching for one's origin is also a pervasive theme of travel, where people reconnect a place and its social system via genealogical connections.

Rooted sense of place and genealogical narrative

Rootedness can be conceived as a feeling of unself-conscious dwelling whereas sense of place is thought of as a certain distance between the self and place that allows the self to appreciate the place (Tuan, 1980). Ideas of roots and ancestral belonging are important in indigenous people's political claims and are one of the bases of ethnic nationalism (Nash, 2002). The Māori, the indigenous Polynesian people of Aotearoa New Zealand for instance, derive their identity from their homeland and connection to place, via which they can maintain strong connections to their ancestors (Tuan, 1977).

Whakapapa, a Māori term, is a fundamental principle in Māori culture. It exists as a genealogical narrative that is told from ancestors to the present day (Te Rito, 2007). It reflects place in terms of its various layers of meaning and memory and is recited in the proper order of genealogies (Williams, 1971). *Whakapapa* is the knowledge and practice of kinship relationship and is fundamental to traditional

Māori identity. As such, Māori are able to trace their lineage as far back as the creation of the universe (Waiti & Awatere, 2019). The core features of traditional social organisation of Māori culture, *whanau* (family), *hapu* (sub-tribe), and *iwi* (tribe), are reflected in *whakapapa* (Lawson-Te Aho & Liu, 2010). Māori use *tangata whenua* (people of the land) to indicate their collective and individual identities to ancestors and places (Te Rito, 2007). *Whakapapa* is manifested in a sense of belonging, purpose, and reality, as well as contributing to a sense of identity in relation to others. This implies that being a Māori is not simply being an independent person, but that you are part of a web of social relationships regulated by the fundamental human–place connection. *Whakapapa*, therefore, perpetuates a value-based approach to relationships where Māori develop their sense of place through connections to both physical and spiritual worlds (Hudson, Ahuriri-Driscoll, Lea, & Lea, 2007). However, it is also important to note that unlike the Western concept of genealogy, *whakapapa* extends beyond people to include environmental features such as mountains, lakes, rivers, and oceans (George, 2010) that are also part of individual identity.

The Māori sense of place is based on their cosmology and culture, which roots them to their tribal territory spiritually and emotionally (Hay, 1998a). Thus, for Māori a feeling for the land is central to their sense of place, particularly, if that land is part of their ancestral tribal territory (Hay, 1998a). There is also a feeling that ancestors (*tūpuna*) are still around, directing their actions (Hay, 1998a). Māori cosmology, therefore, provides strong ties to tribal territories through both myths of origin and connections with spiritual energy (*wairua*), including that of their ancestors (Hay, 1998a). Connections to Māori culture are reaffirmed through the dynamic performance of tribal rituals during community events, such as when recounting tribal genealogy and history through speeches (*whakapapa*) and in action songs (*waiata*) (Hay, 1998a). These cultural practices reflect the *whakapapa* and imbue *māoritang*a (Māori cultural ways). As an example, the *mihimihi* is a welcome speech of the host sharing where they come from and who they are in relation to the place at the beginning of a gathering (*hui*). In a *mihimihi*, the speaker locates him/herself to a place and an ancestry, to reinforce the relationship between this individual and their *whenua* (land), their *whakapapa*, and their *tūpuna* (ancestors) (Hudson et al., 2007). Thus, Māori indigeneity can be understood as a dialogue between the past and the present as well as between location and movement. Other aspects of Māori culture that emanate from *whakapapa* include the sense of place in relation to nature and the environment. This unique relationship between Māori and the environment manifests in a sense of place that reflects a diversity of thoughts and feelings related to spiritual, affective, familial, and physical and cognitive perceptions (Waiti & Awatere, 2019) that is critical to understanding Māori tourism.

Māori identity, tourism, and sense of place

The Tourism Strategy for New Zealand (Ministry of Business, Innovation, and Employment and Department of Conservation, 2019) aims at making Māori culture the heart of New Zealand's tourism offering through the delivery of high-quality and authentic visitor experiences. Organisations such as Tourism New Zealand and NZ Māori Tourism are encouraging, supporting, and promoting Māori tourism products and experiences that are owned, controlled, or delivered by Māori and are focused on Māori values. An important part of the tourism growth strategy is the relationship between the state and Māori tourism enterprises. Māori tourism alone contributed NZ$214 million towards the New Zealand economy in the year ending February 2015 (Puriri & McIntosh, 2019), with this sector expected to grow substantially in the coming years, albeit given the impacts of COVID-19. Although it must be noted that there is a lack of agreement on what constitutes Māori tourism, with terms such as 'Māori tourism business' and 'Māori tourism product' often used interchangeably (McIntosh et al., 2004).

Māori tourism encapsulates many different things, including the level of control by Māori of the business, the nature of the tourism product and extent of cultural content, embeddedness of Māori culture and identity in the business philosophy, and a focus on the unique cultural values of this ethnic group (McIntosh et al., 2004; Puriri & McIntosh, 2019). The questions for Māori tourism development are, therefore, not so much related to definitions of Māori tourism, though important, but more so about the cultural values that should underpin Māori tourism and how these values should be applied in throughout the business, including the development and marketing of Māori tourism products or experiences (Puriri & McIntosh, 2019). At the heart of the business should be *manaaki manuhiri* – care and respect for all people, from all places (NZ Māori Tourism Society, 2020), but also sustaining the integrity of Māori culture by providing benefits back to Māori communities and their culture (Tahana et al., 2000). Māori are increasingly utilising tourism as a vehicle to preserve and promote their culture and create a more prosperous future for the younger generations. This reflects a consideration for a sustainable future that is strongly rooted in economic advancement of Māori, while taking care of people and the environment.

For tourism purposes, there are several values that are important to Māori, which imbue what has been called 'Māori centred tourism' (McIntosh et al., 2004). For example, values of *ōhanga/whairawa* (economic prosperity and wellbeing), *manaakitanga* (shared respect, hospitality, generosity and care for others), *kaitiakitanga* (guardianship/sustainability), *whanaungatanga* (family and belonging), *wairuatanga* (state of being spiritual), *kotahitanga* (unity and solidarity), *purotu* (principle of transparency), and *nga matatini maori* (Māori diversity), among others support the industry's aim for a more sustainable and inclusive tourism (McIntosh et al., 2004; Puriri & McIntosh, 2019). These values are inextricably linked to Māori identity, which emanates from *whenua*, *whakapapa*, and *tūpuna*. There has been several calls for better integration of Māori values and principles that can facilitate Kaupapa-driven Māori tourism businesses (Amoamo & Thompson, 2010; Puriri & McIntosh, 2019) which can also give Māori the confidence to dictate the terms and conditions on what parts of their culture they want to share with the world (Puriri & McIntosh, 2019).

While Māori culture certainly offers a point of difference for tourism in New Zealand, the questions of Māori identity in relation to 'others', 'who' represents for 'what' purpose, and 'how' the culture is represented are contentious ones.

The bi-cultural context of Māori /non- Māori that imbue socio-political relations calls into question the notions of 'being Māori ' (Amoamo, 2011). There is also some cultural unease between Māori identity and that of some European (pākehā) New Zealanders in terms of what constitutes 'Kiwi' identity, especially given the multicultural nature of New Zealand society. However, Oppenheim's (1968, p. 277) comment that 'the culture of the Māori receives only passing attention from the pākehā majority and indeed its very presence as a separate entity within New Zealand culture is sometimes doubted' is now extremely dated. Although some older generations of pākehā may wish that Māori culture could be embedded 'safely' into the European New Zealand society, where being Māori would simply become 'one of us' as New Zealanders (Ryan, 2002), biculturalism is increasingly becoming integral to New Zealand life within a rapidly changing multicultural context. Nevertheless, it does potentially suggest that there may be differing notions of sense of place between Māori and pākehā given the different cultural contexts they come from, although increasingly many pākehā also embrace the land. For example, Māori tour guides welcome visitors and give insights into both the Māori way of being and way of life by retelling myths and legends. These, along with exposure to arts and crafts, potentially further cements the Māori identity as distinct from the other (pākehā). However, as some have argued, there is an omnipresent need to showcase contemporary Māori culture alongside its historical aspects, to present it as a 'living' culture (Amoamo & Thompson, 2010).

Amoamo (2011) suggests that the Māori identity is a 'hybridized' identity because the changing Māori culture is continually affecting tribal identities. There is increasingly a recognition that different *iwi*, *hapu*, or *whanau* have different expressions or interpretations of *tikanga* (customs and traditions) in tourism (McIntosh et al., 2004). In relation to urban tourism products, Māori-owned businesses are constantly integrating tribal and self-identities in the creation of contemporary experiences for tourists (Amoamo, 2011). This is driven by Māori having their own aspirations, within their own terms and values, with regard to their participation not only in economic development, but more specifically in tourism. While the ability for Māori to participate in tourism development has improved over the years (Carr, 2017), for Māori tourism businesses to have cultural legitimacy, the correct ways of *tikanga* must be carefully managed and planned for. *Tikanga* is the active embodiment of the Māori worldview or value system based as it is on relationality or whakapapa (Amoamo et al., 2018). This implies that cultural experiences that contribute to visitors' sense of place must follow the correct protocols, customs and traditions. Without these, there are negative ramifications for perceptions of cultural authenticity by visitors.

Alongside issues of what is Māori tourism and its relationship to Māori identity, there are also issues related to how Māori culture is represented, including what aspects and by whom. In the early days of tourism promotion, Māori identity was framed very narrowly in tourism brochures and marketing communications. For example, several studies confirm that the images of fierce Māori warriors, the *haka* (war dance), the *poi* (women's dance with feathers), and the *wero* (the ritual challenge with grimacing face, tongue poking, and rolling eyes) have shown a high degree of persistency in marketing the tourism experience to international visitors as depicted in postcard pictures and international tourism campaigns (McIntosh et al., 2004; Ryan, 1997; Ryan & Crotts, 1997). Through the analysis of 216 tourism brochures, Olsen (2008) confirmed that Māori are depicted in primarily two settings for tourism purposes, entertainment, and sights. In both settings, Māori culture is portrayed as a timeless traditional people. Such images cater to those seeking the 'exotic', 'primitive', and 'other' (Amoamo & Thompson, 2010). New Zealand's tourism marketing efforts have shaped a stereotypical and homogeneous representation of Māori culture, which has led to a Māori identity that is at odds with the diverse reality that makes up this ethnic group (Amoamo & Thompson, 2010). Previous imagery of Māori culture, thus, has been grounded in a political relationship of unequal power, whereby tourists are exposed to simplistic reconstructions of culture, identity, and therefore place (Amoamo, 2011).

The process of colonisation within New Zealand has meant that, in the past, the promotion of Māori tourism was not controlled by Māori themselves, resulting in promotion of the idea as Māori as exotically different, to be observed as entertainment by tourists (Amoamo & Thompson, 2010) as part of a Romantic portrayal of a South Pacific fantasy idyll invented by Europeans (Hall, 2009). This created a cycle (Hall & Tucker, 2004) where tourists wished to experience aspects of Māori culture, such as the haka, hongi, waiata, and hangi, that were familiar due to decades of marketing promotion and tourists' own experiences of staged cultural demonstrations, then further reinforced the idea of a fixed and unchanging culture. However, there is also a growing demand for Māori tourism experiences that are authentic and provide opportunities for interaction between host communities and visitors. In particular, visitors are looking for experiences that connect them to both historical and contemporary Māori culture (McIntosh et al., 2004). To achieve this, there has been a call by indigenous tourism stakeholders to control the representation and meanings of their image, tribal identities, and cultural difference (Amoamo & Thompson, 2010). As an example, Māori employees can add to the authenticity of Māori tourism experiences. However, it is not always possible to have only Māori employees in tourism businesses owned by Māori. For example, at Blue Penguins Pukekura, a wildlife tour business in Otago, New Zealand, non-Māori employees are included through *manaakitanga* – treated as guests whose training has granted them the right to share stories of the land in ways that are culturally appropriate. By telling stories that explain their own connection to the area as well as highlighting the history of the land and the ways that Māori stewardship of the land is benefiting both the community and the animals in their care, these non-Māori employees have embraced Māori values in delivering the tourist experience, and thus provide authenticity to the experience. By also structuring the business around Māori values and acknowledging the tourists visiting their business as welcomed guests (*manaakitanga*), the tourist experience provided by Blue Penguins Pukekura remains authentic to the Māori ways of doing things. This enables them to balance commercial realities (*ōhanga/whairawa*) alongside their responsibilities to their wider whanau and as guardians of the land and animals (*kaitiakitanga*) (Amoamo et al., 2018). The explanations of Māori history, culture, and the modern interpretation of Māori values provided by employees add value to this wildlife tour, creating a unique and authentic experience for tourists.

The experience of place according to Māori culture is, therefore, strongly related to the concept of *whakapapa* and this influences attachment to people, places, and the environment. Māori attachment to their ancestral lands can be viewed as the root of sense of place, but concepts such as place attachment derive from a broader base of spiritual and genealogical ties as a result of the meanings and history within the place (Waiti & Awatere, 2019), reflecting an interplay between *whakapapa, whānaungatanga* (cultural relationships), and place. Spiritual engagements can generate strong affective connections (Waiti & Awatere, 2019). Sense of place is also sustained through ancestral connections (Waiti & Awatere, 2019). The spiritual presence of ancestors can foster a sense of belonging and attachment to place (Waiti & Awatere, 2019). It is important to note that within a Māori worldview, an ancestor can be either a human ancestor who has passed on, or a spiritual ancestor that may be in the form of an animal or creature (Waiti & Awatere, 2019). Sense of place is not only rooted in cultural concepts of *whakapapa* (genealogy) but also those of *kaitiakitanga* (guardianship) and *wairuatanga* (spirituality), among many others discussed previously. This implies that sense of place is a constantly evolving concept that has both historical and contemporary aspects. As illustrated in the context of tourism businesses, conveying sense of place to visitors through the lens of Māori cultural values involves not only showcasing traditional Māori cultural practices but also the evolving nature of the culture. These values create cognitive, affective, and behavioural ties to place, which subsequently inform attachment.

Box 2.2 *Ngāi Tahu* and connections to place post-tourism development

Ideas of genealogical roots and ancestral belonging provide an important foundation for indigenous people's political claims. Often in the history of migration and colonisation, as well as other cultural flows and transnational and diasporic geographies of movement, indigenous communities have gradually lost some of their connections to their lands and homeplace (Nash, 2002), leaving their genealogical and historical connections to narratives and stories.

Ngāi Tahu, the largest Māori *iwi* (tribal unit) by population in *Te Waipounamu* (the South Island of Aotearoa New Zealand), provides a good example. *Ngāi Tahu* made its home in South Island of New Zealand many years ago, after a long-distance travel by sea from *Hawaiki Nui* (the traditional Māori place of origin) (Te Rūnanga o Ngāi Tahu, 1997). Since arrival, *Ngāi Tahu* formed its permanent and semi-permanent *hapū* ('sub-tribe' or 'clan') settlements in coastal and inland regions of the South Island

and utilised an intricate network of *mahinga kai* (customary food gathering sites) sourcing seafood, eels, birds, and plants and developed traditions, knowledge, and art to guide future generations (Te Rūnanga o Ngāi Tahu, 1997).

One of these important *mahinga kai* is Queenstown, a major international and domestic tourism resort, and its surrounding area. This area is a large and remote region of 8,467 km^2 (Woods, 2011), rich with birds, fish, and greenstone (jade) which is used by Māori to make tools, weapons, and decorative ware. In order to gather these resources for their own use or for trade *Ngāi Tahu* have been travelling between their costal settlements and this region seasonally for centuries. Many *Ngāi Tahu* stories and traditions have developed from this area. However, no *Ngāi Tahu* people have set up any permanent settlement in this region (Ngāi Tahu Tourism, 2020).

As Scottish settlers moved to the South Island of New Zealand in the 1840s (NZ Ministry for Culture and Heritage, n.d.), European and other migrant settlers have gradually moved into this area with ease, ignorant to the established and existing *Ngāi Tahu* traditions, and leading to *Ngāi Tahu* to lose some of its connectivity to the area (Ngāi Tahu Tourism, 2020). As for tourism development, Queenstown has attracted visitors since the 1860s and has had most of its tourism infrastructure developed since World War II (Pearce, 1980). In contrast to the fast tourism and population growth in Queenstown (as well as other tourism destinations in New Zealand) (Woods, 2011), Māori people are economically marginalised in their own land. For instance, in 1991, the unemployment rate in the Māori was over 25%, compared to the national average of approximately 10% (Hall, Mitchell, & Keelan, 1992), while this figure has now dropped substantially, as of the time of writing the Māori unemployment rate (6.7%) remains well above the national unemployment rate (Ministry of Business, Innovation & Employment, 2020). Māori people, *Ngāi Tahu* people in particular, were largely left out from urban and tourism developments in these areas, despite their strong genealogical and historical connections to and sense of ownership of these regions. As a result, Queenstown was developed into a European style tourism destination and resort, namely '*Adventure Capital of the World*', catering mainly to European tourists with little acknowledgement of its Māori heritage.

For Māori communities, tourism offers labour intensive opportunity that does not require high technology or large capital or complex skills to participate, as well as an industry that recognises the value of hospitality essential in individual and tribal Māori life (Hall et al., 1992). In addition, tourism provides *Ngāi Tahu* people to reconnect to areas like Queenstown which are of significant historical importance and strengthen their genealogical bonds to and sense of ownership of these places. Therefore, it is essential for Māori people to take control of the tourism development process.

The Treaty of Waitangi, signed in 1840, is the original agreement of the reciprocal relationship between the Māori and the *pākehā* (European settlers), based upon from which a number of bills and acts were passed to acknowledge and respect *Māoritanga* (Māori culture, traditions, and way of life) (Hall et al., 1992). In the *Ngāi Tahu* settlement of 1998, *Ngāi Tahu* received

> cultural redress in the form of confirmation of the ability for *Ngāi Tahu* to express its traditional *kaitiaki* or guardianship relationship with the environment, tribal redress, an apology from the Crown, acknowledgement of the role of Aoraki (Mt Cook) and economic redress in the form of a payment of $170 million and the ability to purchase property from the Crown (*Ngāi Tahu* Tourism, 2020).

Ngāi Tahu Holdings Ltd. *Ngāi Tahu* Tourism, along with *Ngāi Tahu* Property, *Ngāi Tahu* Capital, *Ngāi Tahu* Seafood, and *Ngāi Tahu* Farming enabled *iwi* to economically and culturally re-establish itself and supported *Ngāi Tahu* people in their cultural, educational, social, and wellbeing initiatives by providing a range of resources and tools (Te Rūnanga o Ngāi Tahu, 1997). Via its owned five businesses in Queenstown region, *Ngāi Tahu* Tourism has reconnected *Ngāi Tahu* community to the Queenstown area and been a significant contributor and decision maker to the local community and tourism development. In addition, *Ngāi Tahu* Tourism has engaged in tourism developments throughout the South Island, most notably in Kaikoura, world famous for its whale watching. Based on its cultural and historical relevance, as well as strong sense of ownership of the place, *Ngāi Tahu* and other Māori communities have become a major player in retaining Māori heritages in the overall development of tourism in the South Island and New Zealand.

Indigenous tourism

Unfortunately in many other parts of the world, many indigenous communities are still marginalised in the tourism development of their own land. For example, participation by indigenous communities in tourism planning and development in many regions of China is difficult, due to lack of experience in relevant tourism fields, entrepreneurship, and management, and sometimes low level of literacy (Wang, Yang, Chen, Yang, & Li, 2010). Under certain circumstances, local government has to lead the initiatives and compromise in balancing the interests of foreign investors with great sources and local's concerns with limited contribution, which often pushes indigenous groups further away from the decision-making process (Yang, Wall, & Smith, 2008). On the other hand, tourism entrepreneurs play a powerful role in ethnic tourism development due to their business and marketing capabilities and resources. However, most are not members of ethnic minorities, and with no sense of belonging to or ownership of the place and thus no sense of responsibility. As a result, these external entrepreneurs exploit minority resources for their financial or business benefits instead (Yang & Wall, 2008). For instance, Wang et al. (2010) found a relatively low level of participation in tourism development by local minority community in Xinjiang, China.

Lack of involvement in local tourism development by indigenous communities may lead or has already led to tensions between locals and visitors, between locals and tourism stakeholders, especially when the destination faces a heritage loss crisis, or an over-tourism problem (Gössling et al., 2020). However, heavy development on tourism disregard of the sources of resources may lead to positive outcomes after all. Xue, Kerstetter, and Hunt (2017) studied identity change among residents of Chongdu Valley, China, a tourism destination with development since mid-1990s, and found changes such as transition from a sense of rural shame to a sense of rural pride and an overall rise in community identity. These changes can be attributed to improved living standards and host–guest interactions, which, if operated well, enhance one's sense of belonging and ownership and attachment to the place. In the case of Hani and Yi indigenous communities in the Honghe Hani Rice Terraces, a UNESCO Cultural Landscape World Heritage Site, Chan, Chen, Ji, Zhang, and Qi (2020) find that local ethnic residents have improved their finance and social status through their involvement in local tourism development.

Conclusions

This chapter has examined several interrelated core concepts such as place identity, sense of place, and place attachment in relation to their contribution to feelings of rootedness and belonging and how this shapes individual and community relationships to place. A central theme of the chapter was how these concepts and relationships can be understood from examining genealogical relationships and the connections that this gives individuals, as well as families and other collectives, to place.

To illustrate these points the chapter examined rootedness from two cultural perspectives: first, from Chinese culture and how Chinese culture has been represented for tourism purposes and what this means for sense of place; second, in terms of how the cultural identity of indigenous populations such as the Māori has been appropriated and represented for tourism purposes and the implications that this has for sense of place. By focusing on the genealogical aspects of sense of place, the chapter has highlighted the similarities and differences that occur in relation to sense of place and place attachment for individuals from different cultures.

3 Stories of place

The narrative sense of place

Introduction

Sense of place, or place attachment, is formed by people sharing culturally emotional meanings to a particular place, providing a combination of individual and collective understanding to the place (Low & Altman, 1992). This understanding implies a long and constant learning about a place via stories, myths, histories, as well as historical, political, social, and fictional accounts (Cross, 2015). This happens in the general context of a resident living and growing in a place, as well as in the context of tourist visitation, in which centring narratives serve as the basis of place and destination marketing (Chronis, 2012a, 2012b; Hsu, Dehuang, & Woodside, 2009; Lichrou, O'Malley, & Patterson, 2008, 2010, 2014). For instance, narratives may be used as an empowering tool to encourage tourists' co-creation behaviour (Pera, 2017), assist in the development of interpretation at heritage sites (Broomhall & Spinks, 2010; Chronis, 2012a, 2012b), as a tool in spatial planning and place promotion (Grenni, Horlings, & Soini, 2020), and are specifically appropriate and relevant in the new digital age (Bassano et al., 2019).

Human society passes on information and knowledge in multiple ways, and social constructionism and discursive psychology suggest narrative as a key in the construction of continuity from past to present, and further projection into the future. Narrative is the telling of stories, including origin myths, family histories, fictional accounts, and political accounts. Phenomenologists take a similar view for their interests in experiences and consciousness and describe sense of place and place attachment as complex, multivalent, dynamic experiences based on the complexity of place (Cross, 2015). Phenomenology is the 'form of philosophy that attempts to give a direct description of the first-person experience' (Casey, 2001, p. 683). For any individual, including a tourist, the phenomenological lived or visited experience is accumulated via each bit of personal experiences with a place, which builds into a self-conscious and perceived sense of belonging to the place (Ferguson, 2011). Narrative is also embedded in a reflection process in an intuitive transition to sense of place, from reading and rewriting to reconstructing a place (historically, psychologically, and geographically), where one may evaluate his/her present relations to the place (Edgerton, 1996). Therefore, sense of place and place attachment can be understood as a phenomenon which is linguistically constructed via interaction in the process of building meanings related to place (Di Masso, Dixon, & Durrheim, 2014). A direct implication of this perspective is that place meanings have an inherently dynamic nature.

Place attachment has an interactional nature and a sense of place changes as its meanings change. A place would not be the same place 20 years after a number of major changes in personal life, or mega social or natural events. For example, the city of Christchurch, New Zealand, went through a magnitude 6.3 earthquake on Tuesday 22 February 2011, killing 185 people and injuring several thousand. This place became fundamentally a different place after this tragic natural disaster, in contrast to the 'garden city' that existed prior to the earthquakes. The stories of sadness, loss, resilience, along with changes in landscapes, are added to the narratives of the city, leading to a modified identity of the place, and a changed image (Hall et al., 2016).

Narratives and stories play a significant role for tourists and visitors in relation to their short and limited actual and physical interactions with a place. An unforgettable story based on a traveller's three-day trip experience to a once-a-lifetime tourism destination may be revisited hundreds of times in this individual's lifetime, with its narratives constantly recreated, and sometimes modified and exaggerated. These narratives are crucial in building the image of this destination, for this particular individual, and surrounding people. One may even say they can be more crucial than the actual experience.

In contrast to environmental psychology's emphasis on individual identity and dependence in the context of studying place attachment, discursive psychologists argue that sense of place is created and modified through continuous physical interactions between people and places, as well as social interactions between people. For instance, Dixon and Durrheim (2000) in a study of apartheid in South Africa found identity change was induced by the narratives that emerged in the sudden transformation of a cherished place. From an environmental psychology perspective, their findings suggest a discursive displacement of place identity in re-conceptualising the person–place relationship. Accordingly, place identity is best understood as a dynamic concept as it can change in response to social and political events and the associated new narratives that may emerge. Narratives in these transitions are also often used for social and rhetorical work, which can lead to ideological transitions that justify sociospatial entitlements, as in the case of post-apartheid South Africa, and change. Hence, this chapter explores the role of narrative in societies in terms of forming a sense of place and its role in tourism development and tourist behaviours, from a variety of perspectives from interactionism, phenomenology, constructivism, and discursive psychology.

Narratives

Human beings are natural storytellers, and narrative is the foundation of human interaction (Cross, 2015). Fairy tales, allegories, myths, and legends are passed on from generation to generation. American academic Walter Fisher, in building the narrative approach to rhetoric and communication theory, named humans as *homo narrans* and specified that the stories we tell held create a meaningful life-world (Fisher, 1984). Narratives serve several different distinct functions in human societies. Firstly, narratives and storytelling are essentially about sharing experiences, which is a common approach for keeping updates of others. These experiences, if derived empathetically, may contain valuable information on improving actions, behaviours, or understandings in responding to the natural and social environment. This information is often organised into knowledge (in the format of success or failure stories) for other individuals or the community (Cross, 2015). Thirdly, narratives represent a way of expressing oneself, stating identities and claiming territories. These functions can be fairly useful in tourism context. For instance, experience sharing may be utilised as an instrument for consumer-to-consumer communication, via stories; narratives may be used to encourage pro-environmental behaviours in tourism context; and loyal customers may be created through tourists' strong identification with a destination or product based on these narratives (Kim & Hall, 2020).

Because of its discursive functions and as it represents a means of building attachment to places, storytelling is especially relevant to human–place relationships (Di Masso et al., 2014). People may reinforce or weaken their bonds to place through the telling and hearing of stories. Accordingly, one's sense of place is dynamic, in terms of what it is, where it is, and when it is. Narratives transform physical land and spaces into a meaningful '*home*' and offer an approach for communicating cultural meanings and social identities (Cruikshank, 1990). Home as

Box 3.1 Digital storytelling in tourism

Digital storytelling refers to 'telling personal stories through digital forms, storing and exchanging those stories in sites and networks that would not exist without the world wide web and which, because of the remediation capacity of digital media, have multiple possibilities for transmission' (Couldry, 2008, p. 374). Digital storytelling therefore refers to online encounter communication practices through technologies (e.g., photography, video, audio) where individuals adopt the roles of storytellers and story receivers in order to build relationship experiences (Pera & Viglia, 2016). Traditional storytelling is focused primarily on listening and reading skills (e.g., oral and written stories), but digital storytelling has a wider focus on other skills, such as interactivity, and it promotes collaboration and literacy through technological engagement. As a result digital storytelling can take advantage of user-contributed content through the traditional processes of selecting a topic, conducting some research, writing a script, and developing an interesting story by combining various types of multimedia, including computer-based graphics, recorded audio, video clips, and music. Digital storytelling has become increasingly used in a number of areas in tourism, including travel and tourism marketing (Bassano et al., 2019), travel and tourism e-commerce (Hassan, 2016), cultural tourism (Wu, 2006), and tourism-related crowdfunding (Kim & Hall, 2020).

well as sense of place, according to Tuan (1991, p. 102), 'is created through naming natural features, classifying them in some manner, and telling stories about them'. One could even argue that human society and civilisations are founded on stories and narratives, and 'shared identity, be it regional or national, is based on the stories that are common to members of the community' (Valdés, 2004, p. 9).

'Humans use stories to create homes, develop communities, recall histories, communicate identities, transmit cultures, and express sense of place' (Huff, 2006, p. 82). Physical places exist in their aesthetic or behavioural possibilities, but sense of place is created by how a place can connect people in society, encourage development of place-related social identities, and reinforce cultural meanings (Shamai, 1991; Stokowski, 2002). Accordingly, language, culture, collectively experienced history, and their derived narratives and stories of a place shape and sustain the cultural and social values associated with place. However, an important point to raise here is that narratives and stories do not have to derive from facts, but can be fictitious. Such fictions have been historically important to place identity, for example, the founding myth of Rome and the Roman Empire is connected to the story of the twins Romulus and Remus being suckled by a she-wolf or John O'Groats the most northerly point in mainland Britain (it's actually Dunnet Head). Nevertheless, landscapes acquires symbolic meanings and resonance once they are embedded in stories and narratives passed on in human society. In tourism context, these symbolic meanings are transformed into destination identity and image.

From a cultural point of view, each culture has its unique origin story, associated with its historical spaces and/or heritages. These cultures also evolve over time, as more stories and histories are built into its narratives. Narratives therefore demonstrate their significance in the dynamics of sense of place, by linking the past and the presence of a place. This dynamic nature of narratives connected to a sense of place, also enables the meanings of a place to be relevant to the people connected to it.

From a social point of view, stories and narratives assist in building the social relationships between people within the place, as well as distinguishing locals (or insiders) from outsiders. These social processes of inclusion and exclusion along with the place creation reinforce individual and social identities and affect the formation of collective identities (Relph, 1976). These narratives are also further shaped and framed by sense of place. This narrative-sense-of-place iterative relationship or feedback loop helps strengthen the preservation of cultures within communities and provides a vehicle for sustaining a culture over generations through stories and oral histories. Sense of place is therefore individual, as well as generational, and is created in the interactions between people and physical places, derived from narratives, and influenced by social interactions. Individuals and society in return feed their experiences and stories back to the meanings of the place which evolve over time, as illustrated in Figure 3.1. However, in spite of its power in the creation of sense of place, the narrative process is only one of several processes, along with other sensory processes. It does not subsume any other process (Cross, 2015). The actual experience of a place, specifically the sounds heard, the colours seen, and the smells smelled, cannot be reduced to any

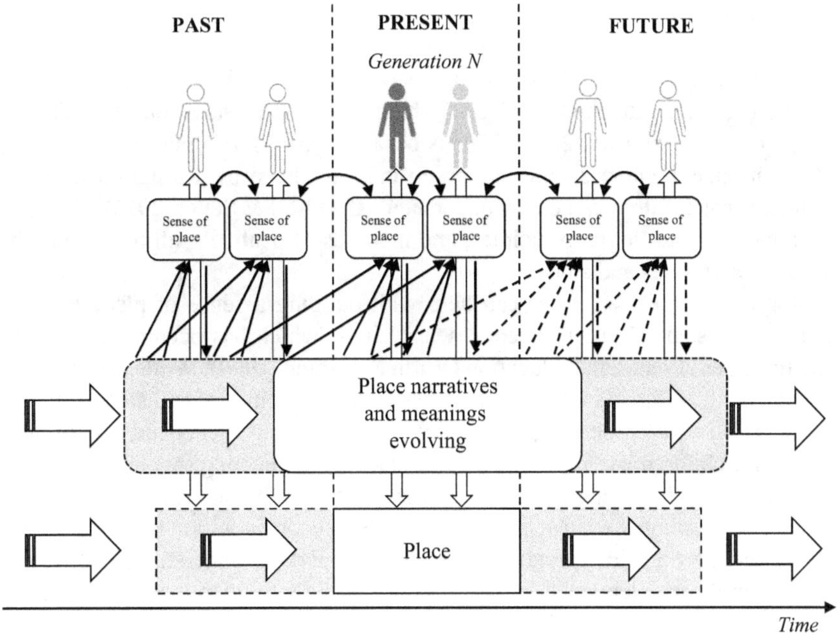

Figure 3.1 Creation and dynamics of sense of place

telling of that experience. Narratives can and do, however, serve to enhance or alter the sense of place associated with that experience.

Homo narrans and their stories of place

According to Walter Fisher (1984), everybody is a *homo narran.* Since being a child we have been surrounded by *homo narrans* and stories that affect how we form and develop our understanding of society, the community(-ies), and the place(s) we live in, as well as how we learn and know about ourselves. In phenomenological literature related to sense of place, scholars have emphasised the importance of social interactions in children's developing sense of place (e.g., Derr, 2002; Hart, 1979; Moore, 1986). For children, stories of their families and homes are told and retold, thereby enabling a sense of belonging to their own families and homes and the places they inhabit. In a wider sense, fables, myths, legends, folk tales, and folklore serve to encourage specific social behaviours, particularly with respect to cooperation among kin, and thus storytelling becomes an important channel for parents to interact with their children and influence their development (Coe, Aiken, & Palmer, 2006). However, at the same time it also serves to connect children to the places that the stories inhabit.

This telling of stories functioning as a means of enhancing children's relatedness to place is also supported in environmental education literature (Lutts, 1985; Sandlos, 1998; Sanger, 1997). From a developmental psychology perspective, stories and narratives provide a psychologically powerful instrument for understanding the environment in which people grow up, as well as the relationships to physical and social space. Storytelling and narratives can help children bond with places in various ways. One way to do this is through illustrating the unique meanings attached to places, which enable it to stand out against others (Blizard & Schuster, 2007). This is because story is a format for organising information, experiences, and standpoints in a way that provides significance and meaning that children can easily understand (Goodsell & Seiter, 2011). Children learn background, logic, principles, and methods of dealing with real life problems via narratives and stories.

Storytelling and narratives can help with building a sense of place through enhancing listeners' environmental sensitivities, which has become a common practice in environmental education (Wirth & Gamon, 1999). While another way of enhancing children's bonds with places is through stimulating listeners' imaginations (Blizard & Schuster, 2007). Through imagining events and occurrences in history, children can make sense of a place by connecting this imagined past to its presence, reflecting their own experiences with the place. In consequence, relevance is established. Such storytelling also becomes important for environmental and heritage interpretation and, although perhaps understudied, is clearly also a significant element in children's and family's tourism experiences (Hibbert, Dickinson, & Curtin, 2013a; Marschall, 2014, 2015).

The power of narratives to engage tourists, including both adults and children, as well as adding a local touch to the tourism destinations, has been widely recognised. For instance, storytelling is used as an instrument to connect visitors to the history of the destination in guided tours at many destinations. Tour guides often tell personal tales that are related to a destination, maintaining tourists' curiosities among loads of impersonal facts and histories of the place (Bryon, 2012). Children can also easily learn the history of an unfamiliar place through engaging its myths and stories. With the development of information technology, multimedia heritage presentation systems that use interactive storytelling further enable visitors to learn and create their own 'stories' and senses of place (Dueholm & Smed, 2014; Nielsen, 2017), as in the case of Saint Laurentius Church in Ename, Belgium (Pletinckx, Silberman, & Callebaut, 2003).

In Māori literature, there are many versions of the origin of New Zealand, in one of which the North Island was transformed from a great fish to a young and brave Māori man Māui caught, while the South Island was formed by Māui and his brothers' canoe. This myth story is a perfect example for driving sense of place not based on real history or facts, but legends and fictions. This narrative not only justifies Māori rootedness in the land of New Zealand but also specifies Māori values and cultural characteristics associated with place. Māori myths and stories have thus been utilised as an educational tool in New Zealand education, as well as a place marketing and branding instrument to educate visitors to New Zealand who have interests in learning about New Zealand's indigenous culture. In addition, a 'pedagogy of place' is developed in cross-cultural education given the significance of narratives and stories related to place, as they enable individuals to situate themselves in interpretive spaces conducive to cross-cultural fluidity and flow (Ferguson, 2011; Giroux & McLaren, 1994). In addition, the environmental education literature also promotes the significant role of sense of place in developing a felt response to place, which is critical to environmental conservation (Wilson, 1997). This is because cognitive understandings on environmental issues alone are not sufficient for motivating care for the environment, unless associated with personal feelings, involvements, and experiences.

As well as narratives and stories acting on children's creation of their sense of place, narrators (or *homo narrans*) actively create the meaning of places themselves through their conversations and interactions with others. Ryden (1993) argued that places do not exist until they are verbalised, via a transformation from thoughts and memories to spoken or written words. Following this logic, a sense of place is created in the common symbolic languages and discourses of people, and can be sustained by rhetorical use of language and narratives (Stokowski, 2002). Narratives and stories represent the meanings of place, but more importantly express these meanings across society.

Box 3.2 Māori myth story of the origin of New Zealand

One of the greatest myth stories of Māori literature explains how the north and south islands of New Zealand were formed. It begins with a brave young man Māui and his brothers setting off on a fishing adventure. Māui dreamed of going fishing with his older brothers. His brothers were reluctant to take him, because he was so young and skinny. Māui came up with a plan to prove himself as a great fisherman, by weaving a strong fishing line from flax as a start. He bound a jawbone, a gift from his ancestor Murirangawhenua (a goddess in Māori mythology), securely to the fishing line. On an early morning, Māui took his fishing line and secretly hid himself in the hull of his brothers' canoe, and did not reveal himself to his brothers until they were out at sea. Māui managed to convince his brothers to keep him in the canoe, and row out to the deepest part of the ocean, where he used his fishing line to catch a great fish. This fish sank close to the underwater house of Tonganui, the grandson of Tangaroa, god of the sea in Māori mythology, and turned into the North Island of New Zealand, best known as Te Ika-a-Māui (the fish of Māui). Māui's canoe, on the other hand, turned into the South Island of New Zealand.

> The stern of Māui's canoe is the southern tip of the South Island, and the prow is the north. When Māui hauled up his great catch he stood on the Kaikōura Peninsula, which is called Te Taumanu-o-te-waka (the thwart or seat of the canoe). Stewart Island is believed to be the anchor.
>
> (Te Ara – the Encyclopedia of New Zealand, n.d.)

See the 100% Pure New Zealand/New Zealand.com website for the role of these stories in tourism promotion and an illustrative map:
https://www.newzealand.com/nz/feature/the-legend-of-new-zealand/

> Just as narrative structures our sense of self and our interactions with others, our sense of place and community is rooted in narration. A person is at home in a place when the place evokes stories, and conversely, stories can serve to create places.
>
> (Johnstone, 1990, p. 5)

Furthermore, in narration, certain parts of *homo narrans'* memories are recalled, a specific emotion or a combination of several emotions are aroused, and thus experiences are to some extent recreated for the narrators to relive that moment. Often a sense of place is reinforced in the recall, reflecting a strong attachment to this place (Chen et al., 2014a, 2014b).

Overall, there are two types of stories contributing in the development of a sense of place: natural history stories and cultural history ones. Natural history stories are found to encourage children to explore the biodiversity of natural places, and therefore motivate environmental protection behaviours, in the prevention of the loss of such experiences (Nabhan & St. Antoine, 1993). Cultural history stories, on the other hand, are essential in strengthening a sense of belonging in the society (Blizard & Schuster, 2007).

Cultural sense of place via narratives and authenticity

In many cultural systems, land is loaded with symbolic meanings, which provides members with both a sense of place and a cultural signature. The intimate relationship between people and place is thus paired with cultural identity and provides people a sense of belonging (Montejo, 2004). Stories in one culture reflect the traits and values of that community and help distinguish this community from another.

Different cultures are reflected in the contents of stories and narratives, and many even regulate the way how stories are told. For instance, oral storytelling and narrative was a common tradition in pre-Conquest and pre-independence Latin America, and is found fundamentally important in building a Mayan sense of place (Huff, 2006). Mayan stories are diverse, centring on maize and the land. This indicates the centrality of agriculture to Mayan existence, which is similar to many other civilisations such as ancient China.

Creation myths are the most symbolic expression of the values and philosophies of a culture. There are two orders of myths: one relating to human nature and the physical world, and another that links people to the wider society (Campbell, 1991). These two orders of myths are essential to a sense of place because myths help provide a resonance to the human experience. In the case of Mayan culture, myths are linked to agricultural values, reflecting the fertilisation of the earth and the harvest of food plants (Huff, 2006). In Mayan myth, humans were made out of yellow and white corn which were harvested from the earth. This was after two failures of making human being (1) out of mud and (2) out of woodcarvings. Interestingly, in ancient Chinese myth, human being was made out of mud by the goddess *Nüwa*, also suggesting a direct origin of human species from earth. These myth stories evolve to reflect their respective culture, such as Mayan culture provides a window into understanding the Mayan sense of place (Huff,

2006), and also become an important element of interpreting Mayan culture and the Mayan archaeological sites in central America to tourists as well as the wider politics of Mayan heritage (McAnany, 2016; Taylor, 2018). Sexton (1999) found narrations in Latin America, known as *cuentos mayas*, are shared.

> … to pass the time after a hard day's work away from home, to keep mourners awake at night during wakes, to exchange information with and entertain a guest or a host, to educate listeners about the values and beliefs of the culture, to commemorate specific historical events, and to entertain both children and adults.
>
> (cited in Huff, 2006, pp. 85–86)

These stories, therefore, feed in the creation of sense of place at both a personal level and community level. Mayan culture is so fascinating and perceived its relevance in sense of place in Latin America that it is demanded and expected by visitors and tourists. This creates bizarre situations where, villagers from formerly Mayan villages adjacent to Mayan ruins who had abandoned this indigenous identity were driven by the tourism demand to reclaim their Mayan identities and cultures (McAnany, 2016; Medina, 2003; Taylor, 2018). However, their essentialised Mayan culture may not be authentic, since many Mayan traditions have long been lost.

This Mayan village case is not alone. Many cultural destinations with historical heritage are facing similar dilemmas. On one hand, they need to address the expectations from visitors towards an authentic cultural experience, but on the other, many cultures have lost parts of their history, or have changed dramatically due to social or political developments. For example, in promoting Sámi destinations, traditional activities, traditional tents or huts, traditional costumes, and reindeer herds are primarily the focus, while the level of industrialisation of the country in which they live is hidden in tourism promotional messages (de Bernardi, 2019). Kazakhstani stakeholders also have to put much effort in defining the dynamic nature of its cultural heritage and the 'old' and 'new' authenticity to attract new visitors (Tiberghien & Xie, 2018).

In the tourism literature, authentic narratives and storytelling are found to strengthen the importance and meaning of visitors' tourism experiences (Engeset & Elvekrok, 2015; Yi, Lin, Jin, & Luo, 2017). Sometimes a desire for 'authentic' cultural experience would result from the visitors' need for a specific cultural sense of an iconic place that overlooks present-day realities, i.e. the Taj Mahal is actually located close to heavy industry and the air quality is poor because of particulates and other matter (Nagar et al., 2020; Singh & Sharma, 2012), as well as the evolving narratives and meanings of these destinations. Often the tourist market calls for a static representation of culture that does not take into account the complex nature of sense of place, negatively impacting tourism attractions and destinations (Hall, 2007). This has led to many discussions on tourism authenticity in academia (Domínguez-Quintero, González-Rodríguez, & Paddison, 2020), and in some cases the sought authenticity can be purely based on fiction. For instance, tourists to South Tyneside, a town in the United Kingdom, were found to perceive the image of the town in terms of the work of popular fiction writer Catherine Cookson (Pocock, 1992). Such media-related tourism is a two-edged sword. On the one hand it can serve to attract visitors and help business development and employment, but on the other it can also create unrealistic expectations of a place (Beeton, 2016), and potentially could even be unfavourable to other industries unless managed effectively (Croy, 2010).

Reframing and reorienting culture to meet market expectations can create tensions between locals and visitors in their interactions, driving cultural conflicts and creating tensions 'between tradition and change, history and modernity, reality and fiction, and cultural and individuality' (Kolar & Zabkar, 2010, p. 653; see also Hall and Seyfi (2020) for examples of this in a Middle Eastern cultural heritage context). An illustration of this conflict can be seen in Figure 3.2. The search by tourists for an 'authentic' cultural sense of place also puts these authenticity seekers in a difficult position. Their obsession with the remote authentic culture distances themselves from the reality of place and leads to confusions and contradictions when they interact with current locals and are exposed to a different lifestyle than what was expected or, at least, much less variance between the visitors' and current local cultures. As a result,

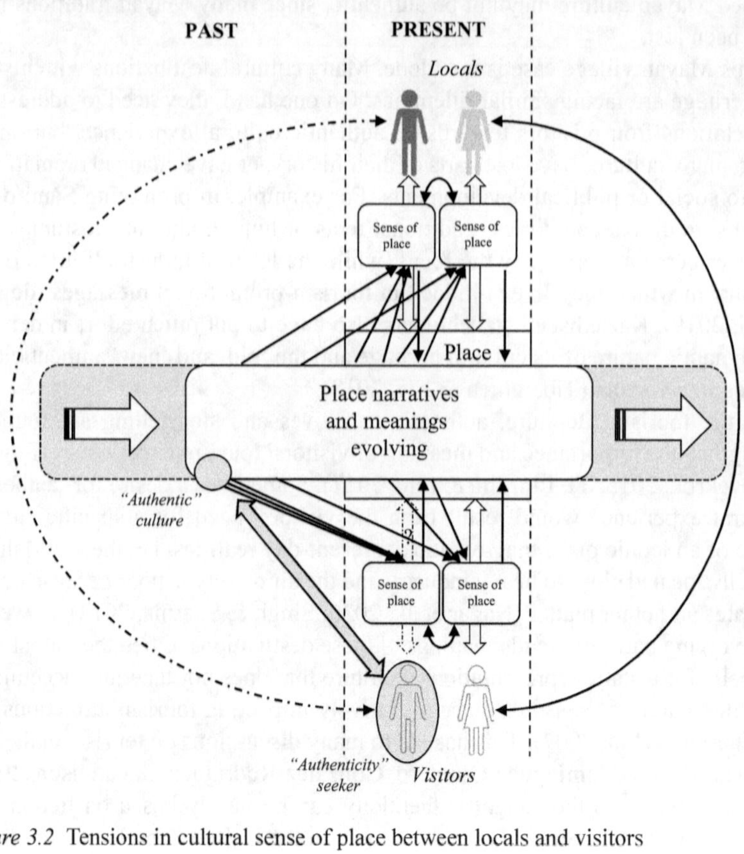

Figure 3.2 Tensions in cultural sense of place between locals and visitors

the visitors' creation of sense of place can be illusional, totally ignoring what is current, and thus compromises their actual interactions with the place. These visitors, in effect, become *Don Quixote*, consciously or subconsciously blocking themselves from their engagement with the current society in the destination. Therefore, such a false sense of place being held by tourists can become extremely disruptive, due to the discrepancies in narratives and stories of the past versus the present.

Generally speaking, storytelling is central in traditional rural and indigenous societies (Sakakibara, 2008). One example is that Inuit traditional legends take part in the establishment of Inuit identity and sense of place (MacDonald, 1998). Indigenous storytelling facilitates the intergenerational transmission of knowledge and experience, allowing for evolution in the form of adaptation to changing environments. Indigenous storytelling and narratives embed facts in values and beliefs (Archibald, 2008), enabling the cultivation of an indigenous sense of place. These stories often integrate knowledge with feelings, which result in emotional connections (Nanson, 2011). Therefore, to create a cultural sense of place in such contexts requires an understanding of indigenous stories in which the spiritual meanings of the indigenous culture are often embedded together with the land. This is specifically relevant to sacred sites in many cultures, where spiritual values are built around nature and cultural identities that are sustained by these places (Lewis & Sheppard, 2005), and which can then be communicated to visitors. However, as Ablett and Dyer (2009) observed, reducing heritage interpretation to the monological transmission of information and targeting universal but individuated cognitive structures is inappropriate. Instead, when dealing with indigenous heritage interpretation, processes need to become more inclusive and culturally situated, and adopt more critically reflexive and dialogical practices.

Box 3.3 Narrative sense of place and movie tourism

Film, television, and literature are important sources for creating a narrative sense of place. As Jewell and McKinnon (2008, p. 160) observed, 'Movie Tourism becomes an additional layer to the ongoing and evolving changes within cultural landscapes and the social constructions and perceptions of society'.

The Chinese *Hugo Award* winning author Cixin Liu's (2017) novel *The Wandering Earth*, which was adapted into a blockbuster movie of the same name, is based in another extreme format of travelling: through space. In the original story of *The Wandering Earth*, the author takes the idea of 'travelling with home' to entirely a different level: travelling to another galaxy taking planet Earth with you (Liu, 2017). This deeply reflects Chinese philosophies in terms of attachment to land and place.

In the movie, an old man talks about when he and his grandson and granddaughter revisited the deserted and frozen Shanghai city and his previous life with his wife in Shanghai, in a story of having a meal together. This plain and simple story of living in Shanghai, a normal place that used to be, demonstrates a great contrast to what it has become, an ice covered series of ruins. His grandson and granddaughter, who have never been to Shanghai before, make sense of the city in the past from this story, while seeing the place under ice. This contrast is intentional by the writer as it helps illustrate how powerless the human species is in facing cosmic level catastrophes. However, in most stories and narratives, the place in which the story sits in is more familiar and realistic so as to relate to and connect with readers.

Dan Brown's (2003) best-selling novel *The Da Vinci Code* was made into a motion picture in 2006. In the novel and the movie, considered controversial, the protagonist, American symbologist Robert Langdon, was summoned to investigate a crime scene displayed like Da Vinci's *Vitruvian Man* (Brown, 2003). Langdon then had an adventure with a police cryptologist Sophie Neveu Saint-Clair in many European cities, and visited a number of historical and art landmarks such as the Louvre in Paris, Winchester cathedral in the UK, and the Church of Saint-Sulpice in Paris (Brown, 2003). In spite of the criticism for using these historical locations and artefacts, as well as harsh criticism and boycotts from several Catholic authorities, the book and movie attract millions of visitors to these landmarks in Europe. Destination marketing and management organisations of Scotland, France, and Britain even organised extensive promotional activities surrounding the film for tourism marketing purposes (Martin-Jones, 2014). The Louvre Museum received 7.3 million visitors in 2005, compared to 6.7 million in 2004, and the administrator Dider Selles attributed part of the boost in visits to the book (Simpson, 2006). In spite of its reality being very different from that portrayed in the book and movie, the Rosslyn Chapel hosted 117,000 visitors in 2005, more than three times the number three years before (Leeman, 2006). In addition, within the narratives of the movie, another main character Sophie Neveu has her own story of searching for her origin in a blood line located in Scotland. Sophie Neveu's early childhood memory in Scotland and her journey in this story result in her returning to her roots, her recuperation of her personal memories, identity, and belonging. Sophie Neveu's personal story itself represents an entire diaspora seeking a return to forgotten, but reconstructed, European roots (Martin-Jones, 2014).

There has been a long history of movies manifesting a narrative of places that lead to people creating a sense of a never-visited-place. This great potential of utilising novels, movies, and other art forms in tourism marketing has been widely explored in tourism literature (Hudson & Ritchie, 2006). For instance, a so called 'Braveheart effect' has been identified, where films, including *Braveheart* (1995), *Rob Roy* (1995), and *Loch Ness* (1996), all had a positive impact on tourist visitation to Scottish locations associated with the movie (Martin-Jones, 2014), something since continued with the *Harry Potter* films and books. However, the Braveheart effect is also interesting because although the film certainly boosted visitation to Stirling in Scotland, an area associated with William Wallace who led the Scots in the late-13th-century First War of Scottish Independence against the English and which already had a 19th-century Wallace Memorial, it was historically inaccurate and much of the movie was filmed in Ireland. So therefore the sense of place that such films convey really centre around imagined rather than real places, although this does not lessen their impact on film locations. If anything, it creates even more issues with respect to meeting visitor expectations (Beeton, 2016).

Another example of film tourism is *The Lord of the Rings* trilogy, which is based in an imaginary Middle Earth (Tolkien, 2012). The movie trilogy helped attract visitors to New Zealand to view the scenery as portrayed in the movies (Jewell & McKinnon, 2008). Specific *The Lord of the Rings* themed attractions and tours were also developed. In spite of the fact that Middle Earth does not exist and New Zealand is not where the stories took place, indeed the Nordic countries were more likely the inspiration, visitors are still aroused in imagination, based on the narratives in the stories and the sceneries in reality (Herbert, 2001). An imagined place in this case creates meaning and provokes emotion responses that are common globally, and assists in fostering a unique narrative sense of places in New Zealand. Since *The Lord of the Rings*, New Zealand has arguably changed because of the cultural significance due to this association, illustrating the power of narrative in creating sense of place.

Tourists' behaviours are known to be influenced by histories, narratives, and representations of potential tourist spaces (Campbell, 2005). People search for images seen elsewhere, sometimes developing into a tourism trip, which represents a 'circle of representation' (Jenkins, 2003). From a broader view, tourists travel to scenes manifested in movies, novels, or on TV, and their travel may be recorded in the creation of new media materials for future tourism marketing or narrative use. In this respect, tourists' somewhat active participation in creating new media products leads to a circulation of media content (Jenkins, 2004). From a tourism perspective, tourists' active participation is a form of co-creation that contributes to the 'circuit of tourism' (Ateljevic, 2000). From a broader place perspective, the visits and their derived narratives can be highly influential in the creation of a sense of place and further the reproduction of space (Larsen, 2006).

Implications for tourism

In a world in which there is an expanding supply of tourism products all claiming to be unique but which often replicates others, stories and narratives can potentially become a unique selling proposition for many destinations (Bryon, 2012).

Stories in tourism engage emotions (Lundqvist, Liljander, Gummerus, & Van Riel, 2013); convey facts and information; enrich, enhance, and infuse facts with meanings (Gabriel, 2000); and thus can attract and maintain people's attention based on their entertainment value (Baumeister & Newman, 1994). Narratives and storytelling are essential to fostering a sense of place in a tourism context for five key reasons.

(1) Narratives and storytelling play a key role in the social construction of place (i.e. tourism destinations);
(2) Tourism is experience-based industry, and thus narratives and storytelling often work closely with physical goods and services in order to create a cohort tourist experience;
(3) Tourism narratives are usually associated with particular memories, customs, values, and stories, in which tourists can relate themselves to;
(4) Through tourism narratives and storytelling, tourists can be personally involved and further participate in co-creation of their experiences; and
(5) Narratives and storytelling seamlessly link facts with fictions, enabling a holistic and congruous information exchange for tourism destinations.

Places are socially constructed into tourism destinations, and narratives play an important part in this process. Via the value-laden, emotion-conferring collective narrative constructions of place, tourism place imaginaries are created (Chronis, 2012). Tourism imaginaries can be theorised 'on four experiential domains: narrative, more valuation, emplaced enactment, and emotional connection' (Chronis, 2012, p. 1797), and narrative is one of the most widely employed discursive devices (Stokowski, 2002). From a learning perspective, one of the most effective ways to know a place is through learning its stories. The notion that tourism experiences relying on the availability and communication of place stories is long established in the tourism literature (Arnould & Price, 1993; Johnstone, 1990). In a way, tourism can be seen as entirely about storytelling (Chronis, 2005, 2008). From a place planning or destination marketing point of view, the meanings of a place are often communicated through marketing materials such as advertisements, commercials, travel books, brochures, and social media (Breek et al. 2018; Coghlan, McLennan, & Moyle, 2017; Young, 1999). On many occasions, the image of a particular destination is largely created on the basis of these marketing materials, rather than other literature and narrative formats, by people who are not familiar with the place, leading to a pure touristic sense of place. Therefore, in one sense, selling a tourism destination is an effort to develop specific place narratives that appeal to different markets (Chronis, 2012).

The tourist experience is a tourism product which is created based on a combination of goods, services, narratives, stories, and human interactions. Of course, tourists may and often do enjoy the food, meals, accommodations, sceneries, and associated services within a tourism destination, simply as consumers. Nevertheless, narratives and storytelling add unique meanings and values to these experiences, creating a sense of cohort and authenticity. For instance, Engeset and Heggem (2015) found storytelling closely works with locally produced food in farm tourism, providing tourists authentic experiences as local farmers. Farmers and winegrowers are able to enhance tourists' experiences by linking the past (via stories and narratives) and present (through food and wine tasting) (Brandth, Haugen, & Kramvig, 2010; Frost et al., 2020). Farmers and winegrowers further become part of the tourism product and experience via their storytelling and related engagement with tourists, demonstration of skills, and delivery of services (Lee & Shin, 2015; Mei, Hågensen, & Kristiansen, 2020). See another example of narratives working with other components of a tourist experience in boxed example 3.4. Narratives add meanings to the consumption of the goods and services, making them unique, and the consumption of these goods and services in return make the stories more realistic, relatable, and unforgettable. Information embedded in stories then interacts with almost all the senses, i.e. vision, taste, smell, hearing, and touch, which are extremely effective in creating a strong sense of place.

Box 3.4 Ko Tane Māori Experience

Ko Tane Māori Experience is a themed attraction centre located in the city of Christchurch, New Zealand, centring Māori culture, wildlife, and conservation to provide guests a real 'Kiwi' experience. A guided tour in this centre includes a premier Māori cultural performance, sightseeing of the native trees and ferns and alongside New Zealand wildlife, and a Hāngī Dinner (Ko Tane, n.d.). In the performance, a traditional Māori powhiri (welcome) is introduced as an official welcome to Aotearoa New Zealand, taking in the wero (warrior challenge) and hongi (pressing of noses between chief and visitor). Before the welcome, guests will be firstly briefed in narratives and stories, and shown the tools and skills of the Māori hunter, his cooking techniques, the games he played to pass away his day, and traditional instruments used to communicate with his gods.

The powhiri is followed by a 45-minute kapahaka performance, which includes songs, poi dances, haka, stick games, and weapon displays. Kapahaka is an expression of Māori culture in song and dance, which displays traditional dance movements accompanied by contemporary or traditional music. Poi in the Māori language means the tethered weights used by the dancers, the choreography itself, and/or the accompanying music. Haka, on the other hand, is a ceremonial challenge commonly associated with traditional battle preparations and providing a formal structure by which to welcome people onto a marae, a communal place, or territory. In the Māori language a marae is tūrangawaewae, a place to stand and belong.

The performance is highly interactive and engaging, each part of which includes three parts: (1) the narratives with stories and explanations of cultural significance; (2) the performance by the Māori performers; and (3) a learning and performance session for the guests to learn poi and the haka, and take the stage performing together with the performers. After the performance and an optional tour to sightsee the native plants wildlife such as kiwis, a flightless bird species only available in New Zealand, guests will be provided with a *hāngī*, the traditional Māori shared dinner with guests which usually consists of a mix of pork, chicken, lamb, kumara (sweet potato), potatoes, cabbage, watercress, salads. and stuffing which have been cooked under hot coals in the ground.

The narratives on these components of the tour, along with some key aspects of Māori culture are important, especially to those not familiar with Māori culture and traditions. Guests will need to learn the rules and procedures in these rituals, such as no smiling or laughing during a haka, and make sense of the cultural significance of these rules and procedures through Māori myths and heritage stories.

Ko Tane Māori Experience: https://www.kotane.co.nz

Different from plain facts and data, stories and narratives work better in being personally relatable to tourists. A story consists of a beginning, a middle, and an end, answering questions of who, when, where, how, and what. This structure prevails among all cultures and all age groups. Even in a context of an alien culture and/or era, one could often easily relate to the point of view of the protagonist in a story. This enables visitors to transform themselves from their ordinary lives to extraordinary experiences (Judd, 1995). Listening to place stories, people usually think narratively rather than argumentatively (Weick, 1995), and establish a discursive sense of place. This approach is particularly helpful for tourists, who have to organise often a substantial amount of unfamiliar information about a destination. Furthermore, in making sense of why and how a destination reaches its present state, one can find stories more moving and convincing (McKee & Fryer, 2003). For instance, in 16th century, the Mughal emperor Shah Jahan built the *Taj Mahal* to house the tomb of his favourite wife, Mumtaz Mahal, out of love, a universal human concept reflecting the feeling of personal attachment, passion, desire, and deep affection that everyone can understand and relate to. Narratives and storytelling can also be specifically useful in understanding complex matters, from historical events to works of art.

Storytelling is a powerful device for encouraging co-creation behaviours in tourism. A proactive tourist who is creative, skilled, and collaborative may engage more in the narratives, and therefore create values to his/her tourist experience, as well as others. In the example of Ko Tane Māori Experience, an engaging guest who responds to the call to the stage, acting as the chief of the visiting group, joins the performance, and becomes a valued and experienced co-creator to other guests. Similarly, all guests participating in the learning and performance session for poi and the haka, as well as all those who join the hāngī dinner (meaning almost everyone), are valued co-creators. Similar to audience who participate in a magic show, an engaging tourist who participates in the narrative or storytelling, may make a huge difference to all other tourists' experiences.

Tourism storytelling also unites the physical world and the world of imaginations for a place. On one hand, people live in a reality perceptible to the senses, which is measured in time and physical distance. On the other, human beings also rely on an internal, sometime spiritual world, that is based on memories from the past, expectations and visions of the future, imaginations, and fantasies (Reijnders, 2016). One's sense of place usually reflects all these dimensions, including the past, the future, the reality, and the fiction. Specifically relevant to tourism context, the destination and its attractions play a supporting role in these fictional stories and narratives, or sometimes are created to fit with the story. For instance, the '*Casa di Giulietta*' or '*Juliet's House*' in Verona, Italy, a popular tourism attraction, is the home of one of Shakespeare's favourite heroines. However, Juliet is a fictional character in the tragedy *Romeo and Juliet*, and of course has never lived in this house. Either way, with the help of these fictions, a form of topophilia, the love of a place, may occur (Tuan, 1974). This sense of place based on fiction or imagination is also referred to as a *dreamscape* (Ehn & Löfgren, 2010), and a *symbolic environment* (Meinig, 1979). New Zealand's mountain ranges, wild rivers, and grassy fields, for example, are famous for its setting the scene for mythical narratives, such as from *The Lord of the Rings* and *The Hobbit* movie trilogies (Buchmann, 2006; Reijnders, 2016).

Conclusions

Besides their roles in fostering visitors' sense of place in the visiting experiences, narratives and storytelling also constantly modify and influence sense of place through these visitors' interactions with others outside the places. For instance, an experienced tourist may tell stories about their trips to their friends and family. To this extent, these visitors become the *homo narrans*, rather than recipients of these narratives. From a marketing point of view, this reflects co-creation in marketing communication. In parallel to various tourism products including tangible goods and intangible services, stories and narratives of a place become another key tourism product of this destination. Story building, storytelling, and story listening represent the production, delivery, and consumption of this product (Bassano et al., 2019). These marketing activities further illustrate the transition from locals' sense of place (identity of place), to tourism product providers' sense of place (value proposition of place), to tourists' sense of place (place image), and finally to potential tourists' sense of place (place reputation). Unlike other aspects of a place that are perceived and evaluated differently, narratives of the place are vital in sustaining a consistent and representative sense of place for all.

Novels, movies, and other forms of narratives clearly drive the creation of a sense of place, especially for those who have had no direct engagement and interactions to that place, which further attract visitations and mobility. These artificial narratives and storytelling contribute significantly in creating new cultural landscapes, interactions between and evolution of society's external and internal economic, cultural and social aspects. From a tourism perspective as part of the development of new understandings of place they also serve to encourage visitation and put a place into the set of potential destinations that a tourist may visit.

4 My land provides me all

The economic sense of place

Introduction

From a biological point of view, places are centres where biological needs including those for food, water, protection are satisfied (Tuan, 1977). A place and its associated community provides residential rights, food, shelter, safety, job opportunities, social connections, and other tangible and intangible resources that are crucial in sustaining one's physiological and social life (McLeod, 2020). From a broader view, this list also extends to the wider landscape and to natural resources (Vaske & Kobrin, 2001). All these resources contribute, ultimately, to people's self-actualisation and self-fulfilment via place experiences (Qian & Zhu, 2014). And, of course, satisfying these needs plays an essential role in fostering sense of place. This process reflects people's reliance on the amenities and resources that places provide, from a functional sense, and a concept of place dependence is defined in relation to this, capturing the conative aspect of one's attachment to a place (Stokols & Shumaker, 1981).

Place dependence can be examined on its own, or by comparison (Stokols & Shumaker, 1981), and develops via the features available in a place (Bonaiuto, Fornara, & Bonnes, 2003). This suggests a theoretical connection between place dependence (reflecting a sense of place and place attachment) and perceived environmental quality and satisfaction. The logic of place dependence focuses on the importance of desirable traits in place attachment and suggests an indication of sense of place for relevant behaviours, including residents' citizenship behaviours such as supporting local products, and support for tourism development and place branding. Indeed, place dependence may be an important element in the development of place promotion and marketing strategies (Hall, 1997). In particular, one's sense of place also largely relies on moral, ethical, and legal residence status. For instance, the fact that an individual can legally reside and live in a country or a place defines the nature of all levels of human–place engagement. The legal and social features of a place play essential roles in shaping one's identity and habits and lifestyle. A specific example lies in food. Geographical and climate characteristics of a place influence food production and consumption, and shape local cuisine into food culture. In some cases place is so important that it serves to define the food product, especially wine and cheese. People make sense of a place through its cuisine and culture, and make judgements on an individual through the food a person likes and dislikes.

Place dependence

Place dependence reflects the functional aspect of attachment to place, based on biophysical aspects of the landscape including natural resources, as well as recreational and open spaces that support social activities (Vaske & Kobrin, 2001). Place dependence considers the behavioural and conative component of peoples' sense of place (Pretty, Chipuer, & Bramston, 2003). Stokols and Shumaker (1981) proposed two angles of examining place dependence in terms of how individuals and communities determine and evaluate a place. The first is to examine quality of place solely and objectively, and the second is to make sense of the relative quality of comparable alternatives. However, making a sense of place and evaluating a place may be relative, depending on whether there is one or many benchmarking places to compare with.

Generic place dependence implies that an individual is attached to a place for functional reasons (White, Virden, & Riper, 2008). Therefore, one may be attached to a place that he/she has never visited (see the discussion on film tourism in Chapter 3), but the place may provide unique resources for achieving certain goals. Functions of places that can satisfy certain needs include employment, education, infrastructure, resource allocation processes, and governance (Relph, 2002; Stedman, 2003). Other biophysical qualities of place that driving dependence include are vegetation, colour, texture, and slope (Carter, Dyer, & Sharma, 2007). Simply from an architectural perspective, a place provides building aesthetics, density, volume; organisation of accessibility and roads including internal practicability and external connections; and ecological and green areas (Bonaiuto et al., 2003). Socio-relational features of place include discretion and civility, security and tolerance, and sociability and cordiality. There are also functional features such as welfare services, e.g. school and social-care services, recreational services such as sport and cultural activities, and commercial services and transport services. Other aspects of a place are also essential such as pace of life, environmental health, and upkeep and care (Bonaiuto et al., 2003). A summary of what a place offers is illustrated in Table 4.1. An individual makes sense of a place and his/her attachment to this place from both aspects mentioned above and a holistic manner. To some extent, this dependence is not for these individuals themselves, for example health care for parents and vulnerable members of society, or education for children. To continue the discussion from Chapter 2, rootedness indicates a profound attachment to homeland, in principle developed from economic or cultural circumstances of the place.

One's sense of place and attachment to place are associated with their perceived environmental quality, or specifically, perceived environmental quality indices (PEQI) (Craik & Zube, 1976). For instance, Chen, Hall, Yu, and Qian (2019) found that residents' evaluation of environmental features and local natural resources interact directly with their sense of place. This is consistent with the logic of consumption of any service product, based upon an evaluation of different dimensions of the service and their qualities, and suggests that basic consumption behaviours such as retention and citizenship behaviours follow this logic (Chen & Dwyer, 2018). Place marketing literature takes this logic

Table 4.1 Aspects of a place that people depend on

Place features	*Areas*	*Sub-areas*
Architectural/town-planning features	*Architectural and town-planning space*	Building aesthetics
		Building density
		Building volume
	Organisation of accessibility and roads	Internal practicability
		External connections
	Green areas	Positive green areas
		Negative green areas
Socio-relational features	*People and social relations*	Discretion and civility
		Security and tolerance
		Sociability and cordiality
Functional features	*Welfare services*	School services
		Social-care services
	Recreational services	Sport services
		Socio-cultural activities
	Commercial services	Commercial services
	Transport services	Transport services
Context features	*Pace of life*	Relaxing versus distressing
		Stimulating versus boring
	Environmental health	Environmental health
	Upkeep and care	Upkeep and care

Source: Adapted from Bonaiuto et al. (2003).

and considers both residents and visitors as consumers while taking a place as a brand or product, and explores how certain behaviours can be encouraged for the development of the place (Kotler & Gertner, 2002). Then topics in contemporary marketing disciplines such as customer engagement, consumer co-creation, and consumer empowerment are also relevant to contemporary sustainable place development. For instance, residents may be considered and examined as a special group of experienced consumers of the tourism products of a destination (Dwyer, Chen, & Lee, 2019).

Given their various roles of locals, consumers, stakeholders, and insiders, residents of a tourism destination may grow a unique sense of place with a strong feeling of ownership and exclusivity, which regulates their perceptions of the outside world and themselves. An intentional distinction between locals and outsiders by local residents is not unusual in tourism destinations, and sometimes it is made even by repeat visitors and second-home owners.

In addition, place dependence implies a mutualistic symbiosis between humankind and the place environment. On one hand, people develop their attachment to place as a result of satisfaction with the residential environment, based on its various attributes. On the other, individuals and groups act towards maintaining the place in a particular manner by providing whatever they need to sustain their lives within the place. This potentially provides an explanation to situations when certain group of locals are against any change made to the place. It also explains, to some extent, local residents' pro-environmental behaviours, illustrating their social responsibilities to their attached place.

In a tourism context, researchers have explored residents' evaluation of benefits from tourism development, which drive their intention to support tourism development in the region. For example, Wang, Shen, and Ye (2020), in an empirical study in Xijiang Miao Village in China, found that residents' place dependence is affected by their perceptions of the economic benefits of tourism development, and further influences how they will support tourism. This finding may help explain the variations in residents' sense of a tourism destination, between those who directly rely on tourism industries and businesses and those who do not. In addition, studies on residents' perceived values of tourism development develop a theoretically similar framework with those on PEQI, outlining benefits from economic, cultural, social, environmental, community, and long-term aspects (Andriotis & Vaughan, 2003; Choi & Sirakaya, 2005; Teye, Sirakaya, & Sönmez, 2002). In contrast to residents in tourism or recreational settings, visitors may also develop their attachment to place for satisfying their specific recreational needs such as hiking (Kyle, Graefe, & Manning, 2005).

As for comparative place dependence, people with multiple place experiences often assess the quality of their place of residence in contrast to others. For instance, Puddifoot (1995) implies comparisons with other places in measuring residents' perceptions of the physical distinctiveness of their physical settings. Lalli (1992) also emphasises the importance of evaluative comparisons in developing a sense of place. Comparative place dependence is particularly relevant in tourism and mobility contexts. The population related to a place cannot be simply divided into a dichotomy between locals and outsiders. There are also long-term visitors, short-term residents, immigrants, returning residents, second-home owners, and so on. These groups usually have rich experiences of having lived in more than one place and which therefore provides comparisons. Based on these comparisons and evaluations, one may develop attachments to multiple places, which are then related to each other. For instance, individuals who had multiple place experiences in their early lives are regarded as being at the risk of confusion in their sense of place and may experience 'cultural homelessness' (Hoersting & Jenkins, 2011). In a tourism context, temporary resident or long-term repeat visitors may play a coordinating role in connecting local communities and visitors. For example, Chen, Dwyer, and Firth (2015) found that Chinese students in Australia balance their attachments to both Australia and China, which further affects their behaviours relevant to tourism.

In summary, people's dependence on place in terms of its resources fundamentally shapes the nature of human–place relationships and interactions, and thus influences the formation of people's identity related to place. For instance, the local community of Grand Manan Island in Canada regards fisheries as the most important local industry for economic and social sustenance, determined by the natural resources available in the place, which dictates locals' social life, behaviour, and island identity (Marshall & Foster, 2002). This identity creation or development may be symbolic in one's social identity, and may also be defined in terms of legal residence status such as citizenship and voter rights, and may also become a contributor to a place's tourism identity (Marshall, 2001; Mair, 2006).

Sense of place and legal residence status

Discussion on sense of place largely depends on the definition and scope of space. Besides looking into place as a geographical concept, scholars also take another view of place as a territorialised political entity (Castles & Davidson, 2000; Jacobson, 1996). From a place dependence point of view, legal residence status implies a means of allocating natural and social resources. For instance, an important system everybody in China faces is the legal residence registry system, also known as the *hukou* system (Qian & Zhu, 2014), which differentiates the allocation of public services (Chan & Zhang, 1999; Cheng & Selden, 1994) and social welfare (Chan, 2009; Wang, 2005; Zhang, 2001).

But even before we discuss the allocation of economic and human capital, legal residence status indicates a basic level of place dependence, the legal entitlement of living and staying in the place, which is bestowed by citizenship. Citizenship links a sense of belonging to political community and offers '*a bounded equality that reflects the historical articulation of national communities*' (Nagel & Staeheli, 2004, p. 5). Citizenship, in the structure of place dependence, services as a legal and cultural framework and a mode of inclusion and incorporation in society (Walby, 1994; Yuval-Davis, 1997). In this respect, citizenship strongly indicates an exclusivity, posing a contradiction to people's individual, abstract, and universal rights (Bauböck, 1994; Kymlicka, 2001). Modern citizenship accordingly draws a tension between inclusion and exclusion which are of profound significance for tourism (Coles, 2008; Coles & Hall, 2006).

Joppke (2007) summarises the development of contemporary form of citizenship in three dimensions: status, rights, and identity. In the status and rights dimension, Joppke (2007) argues the most significant development in the liberalisation of access to citizenship is the removal of racial and gender barriers, and developing a more diversified citizenry along ethnic, racial, and religious lines. Nevertheless, another form of status discrimination prevails, which lies between one's legal status and one's ability to realise societal rights and privileges (Nagel & Staeheli, 2004). For instance, immigrants and some other marginalised community groups share formal citizenship status and related privileges and rights, but often feel excluded from dominant sectors of society (Piper, 1998), not to mention being a part of the dominant representations of society (Mitchell, 1997), e.g. the inclusion of indigenous peoples as part of the contemporary representation of place, women in fundamentalist faith communities, or Black, Asian and Middle Eastern (BAME) individuals in many Western societies. Citizenship therefore can be distinguished between formal citizenship and substantive citizenship.

In addition, in many tourism destinations, the number of visitors is significantly larger than that of local population, which leads to inclusion of temporary legal status in the discussion. For example, Queenstown, New Zealand, receives

3.2 million visitors annually, while the local permanent residential population of the wider Queenstown area is below 130,000 (White, 2019). In these scenarios, visitors, although greater in number, enjoy less privileges and rights, and thus are restricted in their extent to be attached to these places. For many visitors and potential visitors, a legal permission to enter the nation where the target tourism destination is located is often required (Coles & Hall, 2011). From a broader view, the application experience of applying for these legal permissions, such as a visitor or tourist visa, is also part of tourism experience. Obtaining a visa often occurs at the early stage of travelling, suggesting an important role in creating a first impression of the destination. Many examples of legal entry permits and their impact on tourism flow can be found in tourism studies. For instance, Qu and Lam (1997) found an increase in the number of inbound Mainland Chinese tourists by 238,080 a year to Hong Kong, due to a simplified visa control from Mainland China to Hong Kong. In 1988 Hong Kong introduced a simplified visa procedure with South Korea, leading to a 39% increase in tourism visits (Goh & Law, 2002). Non-visa agreements between countries have been recognised as a means to significantly increase visitor numbers and tourism receipts (Balaz, 1995; Coles & Hall, 2011). The two dimensions of status and rights combine to build a path to place and social identity, which is different from previously discussed genealogical or narrative logic. This status–rights–identity cycle also retains a legal uniqueness of creation of sense of place, in spite of globalisation and hyper-mobility.

Box 4.1 Does the Schengen visa application impede UK-living Asian students from visiting?

International students have become a major force driving tourism demand in many countries and regions such as Europe and Australasia. With a strong international and global view, international students travel in the form of youth tourism, but contribute to tourism expenditure more than traditional backpacker travellers, participating more in sightseeing and shopping (King & Gardiner, 2015).

Tourism academics have widely acknowledged the significant link between international education and tourism (Gardiner & Kwek, 2017). Taking the United Kingdom as an example, the number of Chinese students at UK universities had risen to 120,385 by early 2020, before the COVID-19 pandemic, compared to the number of 89,540 back in 2014 (Jeffreys, 2020), which become a key tourism market for entire European region. Studying in the United Kingdom and travelling in continental Europe has evolved into a popular lifestyle for many international students.

However, the United Kingdom is not a member of the Schengen Zone, and thus an international students' UK residence permit does not grant access to the Schengen area. This means that many international students, for example from many Asian countries, have to apply for a Schengen visa for travelling to continental Europe, such as France and Spain.

The Schengen visa application process in the United Kingdom is notorious in international student community for several reasons:

(1) The applicant must book an appointment with the relevant consulate or embassy in London (or Edinburgh or Manchester), which is often limited in availability. On average an applicant can only make an appointment up to a week ahead.
(2) For many applicants who live outside London, they will need to arrange another trip to London either by train or by car, which itself is already a one-day trip. This is because many countries require applicants to apply in person to have their biometric data taken (Consulate of Spain, n.d.).
(3) Many countries use service agencies such as VFS (Visa Facilitation Services), making it difficult for applicants to directly contact a consulate or embassy regarding any questions or emergent requests.
(4) After a formal application lodgement, an applicant would have to wait for a minimum of 15 (sometimes up to 30) days to hear of the final decision as to whether a visa is granted (VFS Global, n.d.).
(5) If a Schengen visa is granted, applicants outside London will have to travel to London again to pick up the passport, or wait for their passports to arrive by post.
(6) For many applicants, only a single-entry visa is granted.

In summary, to obtain a Schengen visitor visa, an international student may have to apply at least one month before the intended visit, and spend up to £200, including costs of travelling for personally submitting their applications and postage. This amount excludes any cost preparing the required documents and materials.

This reality makes it almost impossible for frequent travel to the Schengen area, and thus may impede international students from fully exploring Schengen member states for tourism purposes. This difficult situation also applies to many immigrants to United Kingdom on non-British passports. For those who work and live in the United Kingdom, it is unrealistic to expect a similar tourism and travel lifestyle as those with British or Schengen passports, who are able to travel to continental Europe at short notice. How much opportunity in European tourism has been lost due to this?

Food, cuisine, and sense of place

One fundamental physiological need satisfied directly by place-related resources is for food. The geographical characteristics of a place help cultivate food diversity and shape regional cuisine and food cultures. For example, a coastal region may have traditionally focused on the availability of different types of seafood over the course of a year. Climate is a major driving force in food availability and determines what farm crops can be grown at a specific geographical location, what proteins are available, and the time of year for harvest. There can be substantial variations in place-related foods in a country. For example, although many Westerners automatically associate China with rice, rice is actually the main staple food of southern China, while wheat-based products are the main staple food in the North (Song & Cho, 2017). Similarly, there is a broad contrast between the northwest regions which prefer wheat and local breads and the rice-eating regions of the east and southeast where rice dominates (Chakravarti, 1974). Indeed, perceptions of the association of particular foods with a place may be substantially different from what is actually consumed, with corresponding issues for tourist expectations.

Food consumption and practices are influenced by the regional economy, ethnic differences, and religious restrictions, which are again interrelated to geographical features. The differences in foods of north and south of the Alps in Europe have long been explored in food geography literature, as those between east and west, for economic and historical differences (Grigg, 1995). In northern Europe the seasons have historically not been warm enough and long enough to grow grapes therefore beer and certain spirits were the basic alcoholic beverage, in contrast grape wine could be produced in the warmer areas of southern Europe. Similarly, although many people would associate olive oil with Italian cooking, in northern Italy the climate makes butter much more of a cooking staple. Indeed, France also has such a north–south divide as a result of climatic difference, with olive oil being a feature of Provençal cooking and butter and cream characteristic of Normandy (Thouvenot, 1978).

Climate also plays another important role in shaping food culture. Cuisine styles are often developed as the result of a number of different factors. Spicy food tends to be preferred and has evolved as a part of the mainstream foodways in hot and humid areas, such as Southeast Asia, Southeast China, Latin America, and Western Africa. In cold regions such as northern China and northern Europe, meat stew is a popular cuisine because of the limited vegetable supply in winters and for helping the body cope with cold weather. Once dietary habits and practices are developed they tend to stay with people when they travel. However, they can change over time as a result of migration to a new country, for example, which provides a new food environment and cultures for individuals to engage with. In such cases however, new hybrid foodways may develop that serve to combine the experience of the old and new places as a statement of an individual or community's identity, although acculturation may also occur whereby migrants adopt the local foodways and customs so that they fit into place better (Galegos, 2011). In addition, and which may be more significant for short-term

Table 4.2 A cuisine acceptance and adoption matrix

	Locals	*Visitors*
Local food and cuisine	Local lifestyle	Local food tourism
Food and cuisine from visitors' home culture	Local food diversification	Homeland food dependence

travel activity to different places, many consumers often have a strong desire for familiar food and cuisine, derived from their place dependence, which may provide an explanation for the food neophobia among some tourists when travelling (Chang, Kivela, & Mak, 2010; Cohen & Avieli, 2004).

Any tourism destination needs to have the capacities of catering to its locals and guests. The relationships between place and food have at least two main implications for tourism practices. First, for many destinations there is value in maintaining a signature local food culture that serves to enhance tourists' holistic experience of place. Second, it is also potentially important to ensure that a diverse range of food and cuisine is available at a destination so as to retain the interest of visitors from different cultures when they wish to visit foods and dishes with which they are more familiar. A matrix illustrating locals and visitors' acceptance and adoption of various cuisines is illustrated in Table 4.2.

Local food has found its place in tourism, especially in rural tourism development, for maintenance of natural and cultural heritage, as well as economic development (Hall, Sharples, Mitchell, Macionis, & Cambourne, 2004b). Travellers to a tourism destination are usually interested in local food, which contributes in stimulating local awareness, and assisting in diversification and maintenance of plant and animal variety. Outsiders' awareness and interest in local produce may also enhance community pride and reinforce local identity and culture (Hall, Mitchell, & Sharples, 2004a). In light of current globalisation trend in food supply chain, tourism exhibits as a contributing factor in building strong local food identities and sustainable food systems.

Box 4.2 Wine, tourism, place, and terroir

Wine is often recognised as one of the ultimate expressions of place and regional character. Indeed, artisan cheese and wine tasting are very much the ultimate and intimate experience of place. This is in part because wine and cheese are often directly identified by its geographical origins, e.g. Burgundy, Champagne, Rioja and Comté, Parmigiano Reggiano, Roquefort, and Taleggio, while the development of a global system of geographical indications and designation of wine and food only serve to reinforce this (Gangjee, 2016; Meloni & Swinnen, 2018; Watkin, 2018). A wine or cheese's unique character and flavour rely not only on the grape or milk variety and the skill of the maker but also on the unique and often subtle combination of the characteristics of where it has been grown which, in turn, can impart a distinctly regional flavour. The French use the term *terroir* to describe this phenomenon – although it is a concept which is difficult to translate literally into English, it can be interpreted as referring to the 'almost mystical' combination of all aspects of soil, climate, and landscape present in the designated place and region in which wine or other food product was grown (Halliday, 1998). Terroir 'lies at the heart of the French appellation system, built up by a thousand years of practical experience which has led to a precisely detailed delineation of quality' (Halliday, 1998, p. 28). However, it is important to recognise that it includes not just the physical 'natural' aspects of place but also those elements that have developed as a result of human activity, i.e. the cultural landscape, including how the grapes are grown or what type of milk can be used and the specific traditions and rules that apply to the production of wine or cheese in a particular location.

Relph (1976) has also suggested that tourism is fundamentally about the difference of 'place'. Clearly a region's physical elements combine to define it as a 'place' and contribute to the attractiveness of a destination. Perhaps not surprisingly then, in a seminal work on wine tourism Hall (1996) suggested that there is a significant overlap between the elements of terroir and regional tourism branding in terms of not only the use of geographical designations but also what contributes to people's ideas of a place. Hall (1996, p. 114) argued that 'there is a direct impact on tourism in the identification of wine regions because of the inter-relationships that may exist in the overlap of wine and destination region promotion and the accompanying set of economic and social linkages'.

Hall and Mitchell (2002) suggest that the notion of terroir can be directly applied to the tourism destination, and they proposed the term 'touristic terroir' as descriptor for this specific idea of place, which regards the unique combination of attributes of the tourist region as crucial to the locational or regional flavour of the tourist experience of place, not the attributes themselves. Such attributes may include parts of the natural and/or cultural landscapes, the relative accessibility of the region, and/or the specific infrastructure and resources of a place; however, it is the way in which they combine that is important for the expression of a tourist terroir.

Arguably, nowhere is this more prevalent than in wine regions, where the cultural landscape is an expression of a regional culture and identity and both the wine and tourism industries rely on regional branding for market leverage and promotion (Hall & Mitchell, 2008). So much so that Hall and Macionis (1998, p. 199) suggest that

> Wine and food can therefore be expressive of a regional culture as well as a regional environment. Such a relationship is extremely significant

> for tourism because of the possibilities of utilising wine and the associated vineyard landscape as a means of establishing strong regional identity in the tourism marketplace.

Another term that has been used to describe the particular expression of place that is associated with wine regions is that of a winescape. Peters (1997), who coined the term winescapes to describe the wine landscape of wine regions, suggests that wine regions have a very distinctive agricultural landscape that attracts visitors to wine country. However, in an appropriate reminder of the uniqueness of place and of the food and drinks that are expressions of that place, Peters (1997, p. 4) points out that

> The geography of wine does not end with a landscape. Color, smell, and taste of wine, including judgements about quality, most often stamp a region with its identity… We don't hear many people talk about visiting "beer country", for example – fields of barley and hops are usually grown far from where the brewers ply their trade, and those fields lack the grace and beauty of long rows of well-tended ripe grapes as fall approaches. At the same time, however, though they have common attributes, not all "winescapes" are the same.

Box 4.3 Mobile foods and place: Beef noodle soup and General Tso's chicken

Food is an important factor in fostering sense of place, and often reminds of the taste of 'home'. Beef noodle soup is a Chinese dish, popular in almost every Chinese community around the world, the origin story of which is quite interesting. Known as 'Sichuan-style beef noodles' in its early days, the origins of beef noodle soup can be traced to Lanzhou beef noodles, the first noodle in China invented by the Hui people, a Muslim ethnic group, during the Tang dynasty (Chen, Hall, Yu, & Qian, 2019). In fact, many historians tried but failed to trace the roots of this cuisine to the Chinese province of Sichuan (Li, 2016). During the civil war of China between 1945 and 1949, many mainland Chinese soldiers left mainland China for the island of Taiwan along with Nationalists' retreating to this island (Li, 2016). The soldiers who came from multiple regions of mainland China became veterans who had to change their career to sustain their lives after military service, and many started their own businesses in restaurants. After years of living in Taiwan, veterans and their families missed their homelands in mainland China, but could not revisit those places until the 1980s (Yu, 1997). Driven by this nostalgia for homeland, some Sichuanese veterans created the dish of beef noodle soup in order to reconnect their origins (Li, 2016).

Beef noodle soup immediately became popular in Taiwan and subsequently spread worldwide. However, beef consumption had previously not been widely accepted in mainstream Taiwanese food culture. During the Qing dynasty, the government issued a ban against eating beef, so as to protect cattle for rice-farming, and thus the Taiwanese had been prohibited from including beef in their local cuisine for a long period of time (Chang, 2012). Beef noodle soup was a key driver in changing Taiwanese food culture. Along with many other mainland Chinese dishes it has helped shape contemporary Taiwanese cuisine. For instance, soy milk and sesame cake, a favourite north China breakfast, has also become a favourite breakfast food in Taiwan.

> Taiwanese diaspora interpretations of mainland cuisine, whether legacy or invented, have left their mark not only in Taiwan, but on the world stage. The Shanghai-style dim sum restaurant Din Tai Fung, widely regarded as the world's premier purveyor of xiaolongbao soup dumplings, is a Taiwan-based diaspora enterprise that beats Shanghai at its own game. The restaurant made the New York Times' list of the world's top ten restaurants in 1993, and currently has branches in over ten countries.
>
> (Li, 2016, p. 10)

Besides its great taste, Taiwanese beef noodle soup unites a large community of veterans and their families and helps maintain attachments to their homelands. The younger generation's food preferences are often congruent with the prevailing family eating tradition in Taiwan (Hsu, Lu, & Chen, 2014). Therefore, the veterans' descendants, who might have never had the chance to revisit their parents/grandparents' homelands, make sense of those places through the making of beef noodle soup, its background stories, and of course, its taste as well. Interestingly, this nostalgia regarding reconnecting with Sichuan has, in the case of beef noodle soup, further influenced cuisine development in another place, Taiwan. Beef noodle soup has become a signature dish in Taiwanese cuisine, and is called for when younger generations of Taiwanese travel (Hsu et al., 2014). Accordingly, in a scenario of a Taiwanese moving to another place with a different culture, it is likely that this individual will have nostalgia for Taiwan and reconnect to his/her homeland through food such as beef noodle soup, at which point this introduced dish will potentially take an active role in the evolution of the local food culture in the new region.

Another example of a mobile dish that remains related to place is the notorious 'General Tso's chicken', named after the legendary Qing dynasty

General Zuo Zongtang (a.k.a. Tso Tsung-tang, 左宗棠 in Chinese) (Li, 2016). General Tso's chicken is a wok-cooked chicken dish, claimed to be Humanese cuisine, that is popular in the United States and is often mentioned in many American Chinese cookery books (Dunlop, 2006). In spite of the various accounts of its origins, General Tso's chicken is purely an American–Chinese invention, and is relatively unknown in Hunan itself (Dunlop, 2006). According to the widely accepted real creator of this dish, Chef Peng Chang-kuei (彭长贵), who was born in Hunan and lived in Taiwan for most of his life, General Tso's chicken did not pre-exist in Hunanese cuisine and deep-frying is not a traditional Hunan cooking method (Dunlop, 2006).

After its invention in the 1950s in Taiwan, General Tso's chicken was brought to the United States in 1973, after Peng Chang-kuei opened his second small restaurant in New York on 52th Street. Peng Chang-kuei's innovative cooking had been significantly influential in the Chinese diaspora, and led to the opening of a huge number of Hunanese restaurants (Dunlop, 2006). In 1980 Peng Chang-kuei returned to Changsha, Hunan, for the first time since he left for Taiwan and was reunited with his wife and two children who had remained living in his hometown and opened another restaurant in Hunan in 1990, where he introduced General Tso's chicken. However, this dish was not well received by the Hunanese public as it has little affinity with local tastes. Nevertheless, the international fame of General Tso's chicken meant that it was still publicly reclaimed as part of Hunan culinary culture (Dunlop, 2006).

Sense of place and tourism products

Following on from the above discussion on local food, we can further examine the specific dependence relationship of tourism goods and services to the home place of tourists. For instance, Japanese tourists enjoy tasting famous local foods and beverages to satisfy their travelling experiences, but many Japanese cannot endure more than a few days without Japanese food (Nishiyama, 1996). Similarly, Chang et al. (2010) found that for Chinese tourists food from their homeland contributed to a sense of the 'ontological comfort' of home. Such travellers, in spite of sightseeing and shopping for local products, in effect, take their home place with them when travelling. For many tourists, the sense of their home place dominates their consumption behaviour.

In order to meet the demands of such consumers of tourism product a customised service delivery solution has been suggested to cater to their cultures, habits, and practices. Many tourism marketing scholars have taken this approach. For instance, Lee and Zhao (2003) surveyed Japanese travellers' service preferences in US hotels and concluded that hotels should bring in the Japanese *omiyage* custom (gift-giving) for better catering to Japanese travellers. They also find a mixed preference for green tea and hot water, which is an Asian cultural representation of drinking habits.

As for the tourists, encountering components of their homeland can help build a sense of the 'ontological comfort of home'. When a child travels or goes to school for the first time, he or she may take their favourite toy to help maintain a sense of security. Similarly, travellers to an unfamiliar destination, whether for tourism or business or any other purposes, may apply the same tactic. For instance, many Chinese outbound tourists take instant noodles on their trips because of their 'familiar flavour' (Ma, Ooi, & Hardy, 2018). This phenomenon illustrates a kind of culinary 'cultural baggage' that these Chinese tourists have access to when they need it (Lin, Guia Julve, Xu, & Cui, 2020a). This type of travellers' behaviour reflects a 'home place dependence', with respect to constantly reconnecting to their home place and maintaining a sense of place and a sense of belonging. As discussed above, tourism marketers put great effort to bringing in foreign food culture and cuisine, to cater and satisfy those travellers from that culture. Tourism practitioners should therefore recognise the importance of this 'taking home with me' mentality or 'cultural baggage' idea, which does not suggest any failure or unattractiveness of local tourism products, but which meets the needs of the visitor. Indeed, tourism practitioners need to better understand the significance of tourism souvenir products in reinforcing a traveller's sense of the destination. For example, Qing et al. (2015) note that, on average, Chinese spend more on gift wines than for their own consumption. However, the purchase of some specific wine products, often associated with high status, for gift-giving does not make for a viable cellar door sales experience aimed at the Chinese market at a regional scale (Lee et al., 2016), although if you are one of the few wineries whose products have achieved such a position clearly may be very advantageous.

Souvenirs and mementos are an extremely important opportunity to reinforce destination and place attachment together with a strong emphasis on the relationships that tourism businesses and destinations create with visitors. In the case of wine and food, as well as other experiences of tourist products, research confirms that, in terms of the enduring positive qualities of the economic experience of place, service issues are critically important, but that the significance of this may decline over time, while other aspects of place (notably the setting and atmosphere) appear to take on greater importance. Indeed, Hall and Mitchell (2008) highlight the importance of taking a stage of the travel experience approach in seeking to maximise the strength of the relationship between the visitor and the site/place visit (Table 4.3). This means that given that in food tourism visitation to the site of production is a key component of consumer behaviour, e.g. food markets, farm-gate sales, pick-your-own, providers need to take care in the planning and management of the servicescape of businesses, i.e. place in a narrow sense. For wineries, for example, this most definitely includes the vineyard and winery itself, i.e. the winescape, as well as the cellar door facilities, which appear to also take on more significance in terms of negative recollections of the experience. Not surprisingly, the wine itself remains a core component of the winery experience and, in fact, the negative aspects of the visit relating to the wine diminish with time. Therefore, businesses not only have to provide a good

Table 4.3 Wine tourism, place attachment, and the stages of the travel experience

Stage of travel experience	*Wine experience*	*Opportunities*
Pre-visit to destination/place (Anticipation)	Wine from destination/winery at home, restaurant or wine club	Distribution in main origin areas
	Previous experience at winery/ wine region	Positive on-site experiences (past).
	Previous experience of other wineries	Promotional material that uses place attributes as well as wine attributes
	Promotional material and advertising for winery/wine region	
Travel to place	Wine en route, e.g. at restaurant or on transport	Wine on airlines or major stopping points en route
	Airline inflight channels/ in-flight magazine article on destination that includes wine	Media promotions
Place/destination/ on-site visit	Winescape/Winery experience • Tasting • Education/interpretation • Service • Setting • Activities (e.g. tours) • Food • Atmosphere • Sense of place (gestalt experience)	Positive winery experience
	Wine at hotel, restaurant or café in region	Wine in local hotels, restaurants and cafés.
Travel from place	Wine en route, e.g. at restaurant or on transport	Wine on airlines or major stopping points en route
Post-visit (place memories/ recollection and reminiscence)	Wine from destination/winery at home, restaurant or wine club	Distribution in main origin areas
	Souvenirs and photographs	Availability of souvenirs
	Wine purchased at cellar door or via mail order, website, or retail	Mail-order newsletter and website
Pre-visit to destination	Anticipation of the next visit	

Source: After Hall and Mitchell (2008).

product and high levels of service, but they must pay particular attention to the place setting within which the experience occurs. The identification of reasonably strong levels of enduring satisfaction further reinforces the need for product and service delivery to be included in the planning, development, and management of direct sales venues, such as the cellar door, and the sense of place that the setting provides, particularly given the impact of enduring satisfaction on purchasing behaviour over the long term and word-of-mouth. Importantly, every time that a bottle of wine is opened, or a particular cheese or food item consumed, there is an opportunity to reinforce the economic value of the initial place experience and subsequent attachment.

Conclusion

This chapter has discussed the economic sense of place and its connections to migration, mobility, food, and memory. It initially did this in the context of citizenship and the different privileges that, for many people, provide a reinforcement to place. Indeed, legal residence provides several ways to reinforce place attachment. The chapter then went on to discuss other forms of economic relationship to place in terms of food and tourism products. The chapter discussed how central food is in terms of identity and place attachment. However, it also noted how the place dimensions of food and wine were also expressly used as part of the place marketing, branding, and promotion strategies. In so doing the potential for tourism to take advantage of such place dimensions for the reinforcement of place and hence destination attachment was emphasised.

5 Place and I

Sense of place and place identity

Introduction

Sense of place and place attachment are multidimensional and multidisciplinary concepts with both encapsulating the social and natural meanings of place (Massey & Jess, 1995; Stedman, 2003). Different from other place-related concepts such as place satisfaction, sense of place and place attachment go through personal emotions, experiences, and values, and thus are more personal and linked to the individual (Chen & Dwyer, 2018). A sense of place reflects not only the place but also the person. The powerful meanings and values derived from the history, culture, physical characteristics (such as climate and geography), community, stories, and narratives of a place usually have significant impact on one's behaviours (Carter et al., 2007). These influences constantly interact with people's sense of themselves, and their collective group belonging, and where their sense of place is fostered. In this process, individuals reflect on themselves through their own lenses of the external physical and social world, and build their understandings about themselves.

Their place-based personality and identity are built, via defining and redefining who they are related to the place and in the place, as well as the rules of inclusion and exclusion. Place identity, as discussed in Chapter 2, is determined in creating sense of place in biological, psychological, social, and cultural senses (Proshansky, Fabian, & Kaminoff, 1983). Accordingly, one's sense of place is based upon the outcome of his/her identification with places, and thus reflects an identity aspect of sense of place. As Lengen and Kistemann (2012, p. 1162) observed, such cognition 'includes memories, feelings, attitudes, values, preferences, behavioural concepts and experiences, which are associated with the variety and complexity of the physical setting and define the existence of personhood'.

Place identity strongly reflects the correspondence of self to place. It is developed upon one's sense of place (Jorgensen & Stedman, 2001). Further, as discussed in earlier chapters, sense of place is fostered through physical interactions, narratives, and storytelling, which are easily related personally and which thus drive an identification process with place (Dominy, 2001). As such place identity answers three key questions regarding a sense of place: '*Who am I?*', '*What am I?*', and '*What do I do?*' (Ashforth, Harrison, & Corley, 2008).

Sense of place is also dynamic, meaning that place identity is also fluid and dynamic as well. Place meanings and narratives evolve via the physical settings of a place (as illustrated in Figure 3.1). Accordingly, one's sense of place usually corresponds with such ongoing change. An individual constantly redefines his/her personal and social identity in terms of the surrounding environment and, on many occasions, a person may change or switch his/her place identity as a result of interactions with the natural and social environment. However, in general, the fluidity and dynamics of sense of place and place identity derive in part from wider economic and social geographical processes (Hall & Page, 2014; Massey & Jess, 1995).

Place identity can be 'remote', based on virtual interactions with a place rather than living or visiting the place. As noted in Chapter 3, one's sense of place may be dominantly driven by narratives and storytelling. These narratives can be historical, social, cultural, religious, and ritual, and sometimes are rather powerful in affecting people's sense of place than the actual interactions with the physical landscapes. In this sense, place identity may develop without even visiting the actual place or, in the case of migrants, revisiting for many years, especially given the shift into the Internet-based online information age and global economies and diasporas (Carter et al., 2007) (see Box 5.1). Hence, this chapter explores the role of place identity in reflecting one's sense of place, and discusses several characteristics of place identity from its conceptualisation and theoretical development. The implications of place identity in tourism context are further covered and discussed.

Place identity, place attachment, and sense of place

Before the connections of place identity with sense of place are discussed, an important distinction should be made in defining place identity. According to the definition above, similar to definitions in the majority research in environmental psychology and their applications in tourism fields, place identity is part of individual identity (e.g., Proshansky et al., 1983). However, in place branding literature, place identity usually refers to a different concept reflecting the identity of a place (Hall, 2008b). For instance, Deffner and Metaxas (2010) define place identity of a city as reflections of its historical background and '*the particularities that traditionally characterised that city*' (Deffner & Metaxas, 2010, p. 52). This is rather similar to another concept for a place: place image (Elliot, Papadopoulos, & Kim, 2011). To distinguish this from our definition of place identity, this concept is referred to as *identity of place*.

The conceptualisation of place identity, as defined earlier in this chapter, is theoretically connected to sense of place and place attachment. There are many research streams that conceptualise this concept in different ways, such as in environmental psychology, sociology, social psychology, and humanistic geography. In terms of place identity and sense of place, the majority of these streams specify and subsume place identity under a sense of place. This is in line with how sense of place is defined, and how one's identification with a place works. For example, Chen et al. (2014a) propose that sense of place consists of two perspectives:

Box 5.1 Migration culture, the Internet, and the sense of place of the Pacific diaspora

'Migration culture' is now the norm in the South Pacific island states, driven by external factors, such as international demands for short-term and seasonal labour, and internally by growing youth populations, rapid urbanisation, rural depopulation, and shortages of agricultural land (Connell, 2010). For example, as many Tongans now live in New Zealand as are located on the island archipelago as a whole, with even more extreme emigration trends being seen for Niue, Cook Islands, and Tokelau (Cave & Hall, 2015). As a result, 'home' has become an ambivalent idea for second and third generations now residing in countries around the Pacific Rim (Connell & Brown, 2005). However, advances in ICT have helped maintain family and diasporic networks and place identities.

Approximately two-thirds of Pacific Islanders have mobile phone access (Cave, 2012). Gibson et al. (2010) have shown that access to phones and other ICT affects information about the availability of offshore short-term employment opportunities for Pacific Islanders, as well as improving the communication connectivity that reinforces family networks and relations (Cave & Hall, 2015). Vaka'uta (2012) reported a case study of a family of 76 individuals that indicated that cultural notions of the nurturing of Vā (the practice of nurturing/reaffirming relationships to maintain a sense of place, space, and connectedness) are evident in the online 'activities' (types of exchanges) taking place between members of the same generation and across generations. Family members were networked through Facebook, Skype, email, and mobile phone networks. Members were spread across Fiji, Samoa, Tonga, New Zealand, New Caledonia, Australia, and the USA. The most active online members were aged between 15 and 37 and reflect a generation used to connecting in virtual places but they also represent the next stage in an ongoing layering of enablers of mobility, networks, and flows by which Pacific Island transnational family networks and associated place identities are maintained (Cave & Hall, 2015).

(1) a relationship to place that refers to the different ways in which people relate to places; and (2) place attachment that refers to the degree of the bond with places. Place identity, in addition, is the cognitive dimension of place attachment (Kyle et al., 2005; Lee, Kyle, & Scott, 2012; Williams & Vaske, 2003).

An illustration of how place identity (as of identity of place) is defined in the literature, as well as how it is related to sense of place and place attachment is shown in Table 5.1. However, overall, the specifications of the relationships between these concepts lack clarity, although there are still considerable overlaps between these concepts from different theoretical approaches, in terms of one's cognitive, emotional, conative, or behavioural ties and commitment to a place (Pretty et al., 2003). However, there is a consensus in much of the literature which sees sense of place as a higher level and overarching concept.

Table 5.1 The relationship of place identity to sense of place/place attachment

Definition	*Relationship to sense of place/ place attachment*	*Base theories and disciplines*	*Exemplars*
A dimension of self that defines an individual's personal identity in relation to the physical environment, through a complex pattern of conscious and unconscious ideas, beliefs, preferences, feelings, values, goals, and behavioural tendencies and skills.	A dimension of place attachment	Social identity theory; identity process theory; place-identity theory	Proshansky et al. (1983); Twigger-Ross, Bonaiuto, and Breakwell (2003); Twigger-Ross and Uzzell (1996)
	A dimension of sense of place; parallel to place attachment		Jorgensen and Stedman (2001)
	A dimension of place bond; parallel to place attachment		Nielsen-Pincus, Hall, Force, and Wulfhorst (2010)
	A place-related concept including place attachment as a dimension		Cuba and Hummon (1993); Lalli (1992)
Representations of symbolic qualities that include a place's name, its geographical entity, its natural and social histories, its derived social meanings, and other particularities that traditionally characterise that specific place.	A source upon which one's sense of place is created	Place branding/ marketing	Hall (1997); Kalandides (2011); Ramsay (2003); Relph (2007a, 2007b); Uzzell (1996)

Overall, four positions can be identified in terms of the relationship between place identity and place attachment:

(1) Place identity and place attachment can be treated as similar or even the same, and used synonymously (Brown & Werner, 1985; Stedman, 2002). In this sense, both concepts generally reflect the sense of belonging, similar to the meaning of sense of place;
(2) Place attachment is considered as a dimension of place identity, to indicate an affective aspect of one's personal identity related to a place (Lalli, 1992). For instance, Cuba and Hummon (1993) consider both emotional ties and affiliation with place as aspects of place identity, implying that place attachment is a dimension of place identity;

(3) In contrast, place identity is conceptualised as a dimension of place attachment to indicate a cognitive aspect of this human–place bond (Kyle et al., 2005; Lee et al., 2012; Williams & Vaske, 2003). This may be considered the most widely accepted conceptualisation model to this date; and
(4) Along with other concepts such as place dependence, place identity and place attachment are two separated dimensions of a supra-ordered concept, such as sense of place (Hay, 1998b; Jorgensen & Stedman, 2001). For instance, Hernández, Martín, Ruiz, and del Carmen Hidalgo (2010) suggest place attachment as an affective-emotional bond with residence places, while place identity is a cognitive component of self-concept and/or of personal identity in relation to the place one belongs to. According to this proposition, place identity and attachment are two parallel dimensions of sense of place.

These four schools of thoughts are summarised in Table 5.2, although in much of the sociological literature, place attachment and place identity are not clearly distinguished at all (Goudy, 1990; O'Brien, Hassinger, & Dershem, 1994).

A key reason of this lack of clarity between these concepts and their relations is their often high interrelationship. For any individual who has long lived in a place and develops strong sense of place, it is reasonable to expect both a high level of place attachment and place identity and, sometimes, a high level of satisfaction and social bond (Brown, Perkins, & Brown, 2003; Hernández et al., 2010). If place attachment is defined purely affectively (and thus referred to as affective attachment for distinction), its relationship to place identity may vary in contexts as illustrated in Table 5.3. For example, residents are often found with high levels of both place identity and place attachment, and thus there is a high level of correlation between the two concepts in many studies. In a tourism context, this also applies to repeat and experienced visitors such as repeat visit hikers loyal to Baxter State Park in Maine (Kyle et al., 2005). Refugees from a place experiencing social or natural disasters may retain a strong identity related to the place, but find it difficult to emotionally bond to a place that has been destroyed (Williams, 2009). Many tourism studies have found it difficult to distinguish between constructs such as place identity and attachment, suggesting often both high/low performances in the scores of these place-related constructs (e.g., Chen et al., 2018; Lee et al., 2012). It is straightforward to argue both low place identity and attachment for short-term (sometimes once-a-lifetime) tourists, since these two often take time and engagement to develop. Another example lies in Hay's (1998a) study on analysing sense of place in several types of samples including long-term residents, tourists, outmigrants, repeat visitors, holiday home owners, and so on. However, this high correlation issue does limit the application of complex theoretical framework in investigating multiple aspects of human–place relationships.

Place identity and its functions

Place identity symbolises environmental meanings in the interpretation of self. As noted above, it helps answers the questions of '*Who am I?*', '*Where am I?*', and more fundamentally, '*Where do I belong?*' (Cuba & Hummon, 1993). These three

Table 5.2 Four schools of thoughts regarding the relationship of place identity to place attachment

No.	*Conceptualisation model*	*References*
(1)		Brown and Werner (1985); Stedman (2002)
(2)	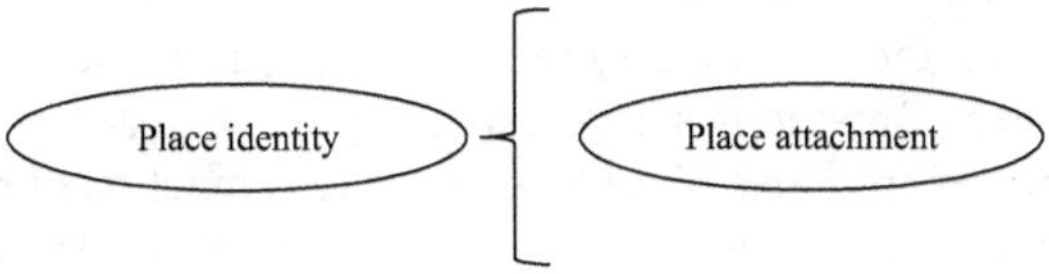	Cuba and Hummon (1993); Lalli (1992)
(3)	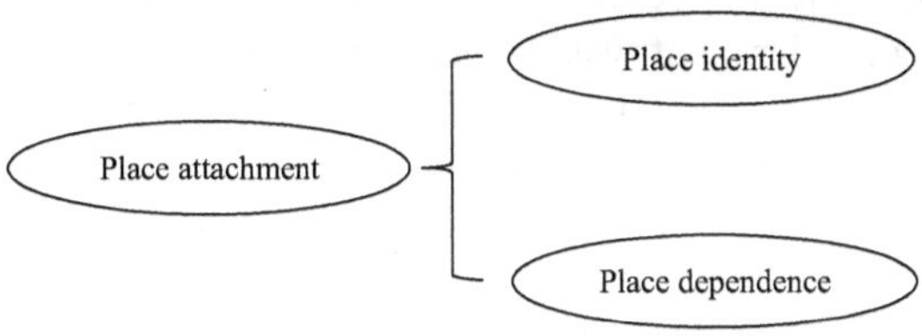	Kyle et al. (2005); Lee et al. (2012); Williams and Vaske (2003)
(4)	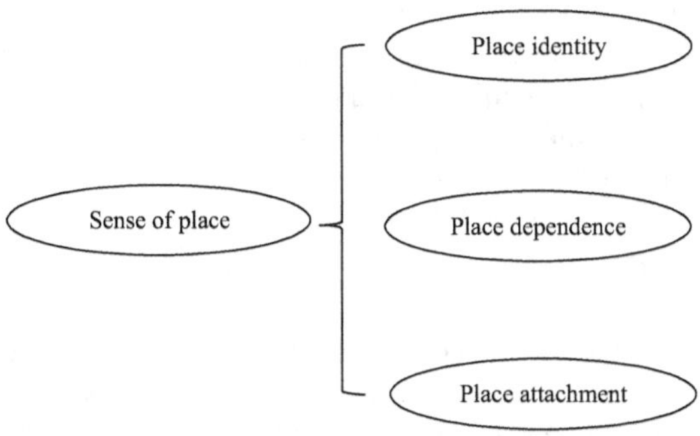	Hay (1998b); Hernández et al. (2010); Jorgensen and Stedman (2001)

Table 5.3 Examples of contextual differences between place identity and affective attachment

Examples		*Affective attachment*	
		Low	*High*
Place identity	High	Refugees/outmigrants	Long-term residents & repeat visitors
	Low	Short-term visitors	Short-term satisfied tourists

questions link one to a physical environment and suggest a relevance between an individual and a social entity. Place identity is a complex concept that needs to be specified in both self-conception and scope of place. It is an indication of a sense of belonging, as well as a sense of self, and ranges in scale from dwelling places to neighbourhoods, communities, regions/states, and even nations (Cuba & Hummon, 1993).

Following the cognitive self-concept definition, place identity can be specified with having five functions (Proshansky et al., 1983). First, place identity has a *recognition* function, which provides an 'environmental past'. This serves to confirm one's continuity from the past, helping to recall previous memories and experiences via the attributes of previous places, especially relevant and significant when an individual moves from one place to another. This function of place identity may negatively drive a dissonant feeling with the current environment and create a dysfunction of the sense of self. In a tourism context, nevertheless, this dissonance often echoes the motive of escape from routine life and enables tourists to act in a desired outsider identity and behave in a different fashion.

Second, place identity has a *meaning* function. Every place has its meanings and purposes which regulate what should happen in it. One's place identity is a source of these meanings, and a group of cognitions linking the past and present. This enables a person to recognise the physical settings, as well as understand their purposes. For instance, Dominy (2001) finds that farmers in New Zealand install meanings of a place into the formation and maintenance of self-identity. In a tourism context, one could quickly create a sense of a hiking resort, and link this meaning and function of that resort to a personal hobby, and further build this place into part of their personal identity. Through this process, a visitor transforms from a repeat tourist/consumer to an experienced new 'local'.

The recognition and meaning functions of place identity build the foundation of making sense of a place and often lead to the function of *expressing oneself*. For instance, Ting-Toomey (2015) in her work on identity negotiation theory states that place identity is a common tool for communicating the self in particular interaction situations within a cultural context. From the example above, this means that the repeat tourist and consumer of a hiking resort can utilise the place as an instrument to meet a specific need that corresponds to their specific tastes.

As emphasised throughout this book, places are constantly changing and thus people need to mediate changes to cope with new realities. This is fulfilled by a mediating function of *place identity* with respect to the skills of environmental understanding, competence, and control. This means that one is able to detect changes in a familiar environment and grasp what needs to be done to mediate the person–environment discrepancies, as a 'local', based on a strong identification of the place. Following this logic, the relevance between place identity and behaviours such as pro-environmental behaviours can be established (Dresner, Handelman, Braun, & Rollwagen-Bollens, 2015; Scannell & Gifford, 2010).

The last function is *anxiety and defence* function. Place identity serves to define, maintain, and protect one's self-identity in a spatial setting. It can also signal threat or danger that may accommodate any anxiety and defence. For instance, if someone feels uncomfortable in a certain physical environment, he or she may defend via fantasies, withdrawal, and even hallucinations in order to escape from reality, through previous memories and experiences. This is especially relevant in the context of immigrants' identity (Liu, 2015).

Through these functions, place identity clearly illustrates one's sense of belonging to a place, as well as meanings of place to one's life. In fact, place identity indicates the great personal significance of a place that often has specific meanings to an individual; this could include genealogical and/or cultural origin, religious rootedness, and relevance. Place identity can therefore help us define who we are, as well as who we want to be, or who we want others to perceive us as. Indeed, think of one of the first questions people ask when they meet someone new: *Where are you from*?

Dynamic and fluid place identity

Both sense of place and place identity have a dynamic nature (Carter et al., 2007). Place identity is a complicated psychological concept determined by both enduring components and far less stable properties of a place. Place identities are subject to change, even for people who have resided in a location for a long time, due to the constant changes that occur in a place. Therefore, a modified sense of place can lead to changes to one's place identity, and place identity, in turn, will affect how one makes sense of a place in the future. This is consistent with the key points in the identity process theory, where identity is viewed as a dynamic concept, created in and modified by social interactions (Breakwell, 1983, 1986; Speller, Lyons, & Twigger-Ross, 2002).

Meanings are not static in time or space, making it impossible to achieve a fixed sense of place. The dynamic nature of sense of place was briefly examined in Chapter 3 and will be further explored in Chapter 7. An illustration of how place identity evolves along with a changing sense of place and a changing place is given in Figure 5.1.

Many factors may ultimately lead to the changes in sense of place and thus place identity. For instance, the physical setting of a place may change, according to urban development and regional planning. Christchurch, New Zealand, was hit by a magnitude 6.3 earthquake in 2011, resulting in a long-term recovery and rebuilding process of the city. After ten years the city has developed into something entirely different from what it was before the earthquake. Perhaps because the city centre is difficult to recognise by its own community as a result of so many landmark buildings being lost people seem to feel uncertain about their place identity. A survey on locals' attitudes found that people are with a feeling that something is missing, in spite of their city pride (Mccrone, 2019).

Besides the physical and tangible aspects of a place, the narratives and meanings of place may change too, as discussed in Chapter 3. In the case of Christchurch, the narrative of the city turned into something fundamentally different, in contrast to its reputation of 'garden city' prior to the earthquakes. The stories of sadness, loss, resilience further led to a modified identity of the place, and further place identity of who live in this city. Social changes may also result in radical changes

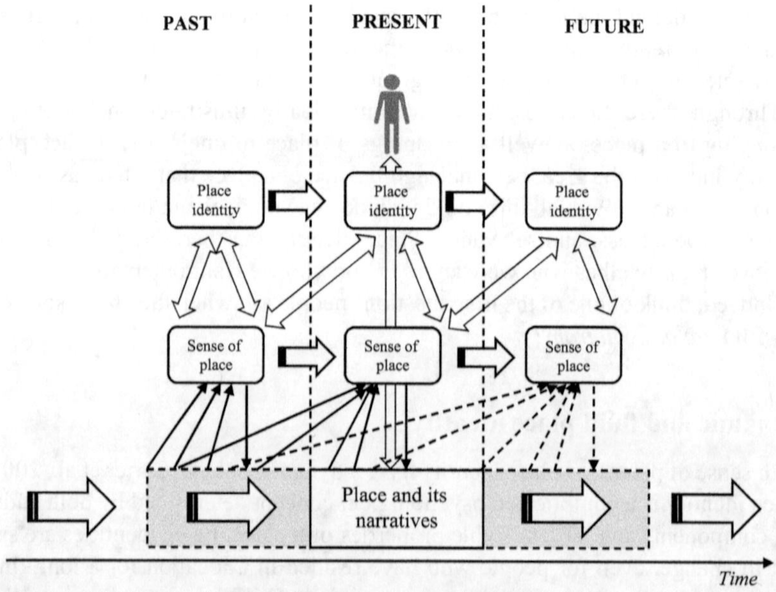

Figure 5.1 Dynamics of sense of place and place identity

in sense of place, which might include political regime changes, demographic changes, cultural changes, economic changes, and so on.

In 1968 a classic sci-fi movie *Planet of the Apes* was released. It was based on French author Pierre Boulle's 1963 novel *La Planète des singes* and was a smash success with both critics and audiences, taking in an estimated $22 million in North America (Greene, 1996). This film tells an adventure story of an astronaut crew on a supposedly alien planet where apes are the dominant species. In the end of this film, it reveals that the planet is in fact Earth in the distant future in the aftermath of a nuclear war. The ending, one of the greatest twists in movie history, is powerful for creating this significant dissonance in recognising something familiar while everything else is so not. It not only changed the protagonist's sense of Earth but also questioned his *Terran* (or earthling) identity, and further challenged the audience's self-identity as the lords of this planet. This ending is relatable to many who have left a place that they are familiar with for rather a long time, and returned as the same person but a felt outsider and stranger for many aspects due to changes. Something that is often familiar for international students when they return 'home' after their study abroad as well as for expatriate workers. This conflict between 'knowing it is the place' (cognitive identity) and 'feeling it is not' (sense of belonging) can significantly affect how one sees and identifies him/herself as well as others. The implications of such conflicts can be so substantial that in the case of students studying abroad, they may even move back to the sojourning country after having come 'home'. Christofi and Thompson (2007), for example, describe the structure of the experience of individuals who returned home after studying abroad, became disillusioned with their home country, and returned to their sojourn country. They found that in the case of such students there were emerging bipolar themes of conflict/peace, reality/idealisation, freedom/restriction, changing/static, and comfort/discomfort which were grounded in the theme of cultural comparison

Box 5.2 Zhizhang He

In China, almost every child learns to memorise a poem from the Tang Dynasty, Zhizhang He (贺知章), describing a scene when a poet came back to his hometown after leaving it for decades. This poem writes and translates:

少小离家老大回
Leaving hometown as a little child, as an old man I came back.
乡音无改鬓毛衰
No changes in my accent, but my hair is no longer black.
儿童相见不相识
A child does not know me, in curiosity he came to check.
笑问客从何处来

He took me as a visitor, not the old homecoming Jack.

In this and some other poems, this poet describes his confusions in his relationship with his hometown, as well as his sense of his hometown. Having seen generational changes in the local community, the poet sees his hometown as a different and unfamiliar place, in spite of mostly unchanged landscapes, and *vice versa*: the place sees him as a stranger too. His place identity has been modified in the interactions with the younger generation of the place, although other personal characteristics such as accent remain the same.

However, one cannot be isolated from society when defining and identifying yourself. Many characteristics in defining oneself are relative terms, such as 'tall', 'intelligent', 'fast', 'quiet', etc. For instance, someone self-defined as 'confident' in Western culture may be perceived and identified as 'arrogant' in certain Eastern cultures. To this extent, place identity helps stabilise one's self-identity in an unfamiliar cultural and/or social context.

For instance, in ancient China, the identity and image of a place was rather static, and mobility was not frequent. China has been mainly an agricultural civilisation with the majority of people strongly attached to the land (Lee, 1978), unless pressured by population increase (Pei, Zhang, & Lee, 2016). After entering a different place, people often introduced themselves in the order of places of origin first, followed by names. For instance, a famous military general who lived during the late Eastern Han dynasty, Zhao Yun, always introduced himself as *Changshan, Zhao Zilong* (Besio, 2012). Here, *Zilong* is his courtesy name, while *Changshan* is Zhao's hometown, reassuring his place identity. Through exhibiting one's association and affiliation with a place, an individual's personality and virtues can be implied, avoiding potential confusions in relatively describing one's social and physical image. Similarly, national identities in contemporary world have widely been applied to indicate collective cultural values and personalities (Hussain, 2019; Neill & Waring, 2019).

The relative nature of place identity is especially relevant in the contexts of migration and ethnicity. Immigrants' personal and social identities are refined by the circumstances of the places they belong to and from where particular beliefs and values function as boundary markers (Brockhall & Liu, 2011). These markers provide instruments to define and determine situations of inclusion and exclusion. Everyone may have to go through an identity reconfirmation or rejection, or a bicultural or multicultural identity construction, in order to adapt to new physical, cultural, and social environments. In reality, and especially in a mobile world, this is often a continuous process of interacting with the sense of place as part of the process of cultural acceptance and coherence (Huang, 2011). For example, in the context of cultural identity, including ethnicity, a specific language, culture, religion, nationality, and a shared heritage are often retained in order to maintain a mode of identity and a symbol of belonging (Song, 2003). Whether on purpose or not, an individual's place identity may be reinforced by external factors, such as the social environment or even seeing a programme on television, that remind them of their origin and sense of belonging. For example, an immigrant usually faces awkward social situations, such as being asked the question '*where are you from?*'. After answering with the place this individual currently resides in, a follow-up question is often addressed as: '*Where are you originally from?*' This sort of social interaction reminds migrants of their ethnicities and their different belongings, but especially their place identities associated with other places. Therefore, from this logic, one's place identity is fluid, not only in time but also in terms of social contexts.

As for people who have lived in multiple places, they very much likely develop multiple place identities, or have no clear place identity at all. Hoersting and Jenkins (2011) find evidence that individuals who have encountered multiple cultural frameworks in their early lives are at the risk of 'cultural homelessness'. Mobility at a later age, on the other hand, does not necessarily undermine place identity, since people are generally able to personalise a place to enable the construction of a coherent sense of personal place identity (Rowles, 1983). One could thus say that for some people mobility between communities and regions may even enhance the identifications with those places.

Box 5.3 Multiple place identities in tourism

Visiting friends and relations/relatives (VFR) is a significant tourism market for many destinations, including Australia and New Zealand. The number of VFR travellers to Australia has been resilient for many years, standing at 2.5 million visitors in the year ending March 2020 and accounting for over 30% of all international visitors and AU$7.3 billion of expenditure (Tourism Research Australia, 2020). The hosts of these VFR travellers are primarily migrants and international students who study in Australia with the VFR travel generated by international students having been explored in several studies (e.g., Chen et al., 2015; Kashiwagi, Nagai, & Furutani, 2018).

The experiences of immigrants and international students well illustrate the nature of multiple and dynamic place identities (Hall & Williams, 2002). Take Chinese students studying in Australia for example (Chen et al., 2015). They may hang out with their friends at local amenities as 'real Aussies' and manifest their international identities whenever necessary, for instance, in classes. When hosting VFRs, they will usually take their homeland identities in the social interactions with these VFRs, while switch their identities to locals in tourism encounters, such as negotiating a lower price as a local. Language use can be a useful tool to detect these identity switches, but does not necessarily always correspond.

Chen et al. (2015) surveyed a number of Chinese students from all over Australia and found that, on average, Chinese students recognise themselves more as Chinese than Australians in general. The lowest mean score on place identity of China is 5.46 out of 7, while the highest mean score on place identity of Australia is 4.79 out of 7. All mean scores are higher than 4, suggesting a dual place identity on average. The structural model testing concludes that the more attached they are to Australia, the less Chinese students are willing to help in a tourism context, while attachment to China demonstrates an opposite effect on willingness to help. In general, Chinese students in Australia illustrate a positive attitude towards hosting their friends and relatives from China visiting Australia and do not mind being reminded of their China identities. On the other hand, a considerable portion of Chinese students have long-term plans of staying in Australia as migrants. Therefore, these Chinese students exhibit an explicit intention of becoming 'real Aussies'.

This multiple and fluid place identity *status quo* can be beneficial, allowing and enabling these individuals to better adapt themselves in various social scenarios. For example, in a local social context in Australia, they may behave totally as locals but contribute to any conversation with their perspective as an outsider. Alternatively, they may take advantages of their ethnic identities in showing a sense of humour when necessary. In the hosting VFR context, Chinese students' Chinese identities can easily comfort their guests and reduce their stress level, in spite of these guests often facing totally unfamiliar exotic natural and social environments. However, whenever required, these Chinese students can utilise their local knowledge and identity to help reduce social distance with locals.

Be that as it may, multiple place identity may bring in certain challenges. A major issue is that the outsider identity may attract racism attacks. For instance, the 2020 outbreak of COVID-19 has generated global conversations in which racist discourses are shaped and perpetuated towards certain groups such as those perceived to be of Chinese ethnicity and outbound Chinese students (Yu, Li, Yu, He, & Zhou, 2020). In Christchurch, New Zealand, a Chinese photographer was assaulted in public on the first day of COVID-19 Alert Level 3 (Sherwood, 2020). This individual has lived in New Zealand for many years, who considers himself as a kiwi, but was believed to be assaulted racially motivated. Moreover, since early 2020, Chinese travellers' mental well-being has been threatened and compromised to an extent, due to potential verbal and physical attacks due to their place and racial identities (Zheng, Goh, & Wen, 2020). However, it is important to recognise that such issues are often contextual and represent the problems ethnic and student minorities face in many countries. For example, African migrants and students faced widespread racism over COVID-19 in China (Pai, 2020). Indeed, what is remarkable with COVID-19 is the way that it has exposed the history of racism and 'cleanliness' in a number of countries in terms of attitudes towards ethnic minorities, and while COVID-19 may be new, 'xenophobia has been intertwined with public health discourse for a very long time' (Burton, 2020).

Place identity in tourism

Upon understanding place identity, its nature, and its complexity, it is worthy investigating its role in tourism context. From the residents' point of view, place identity is important in creating a sense of being insiders and exclusivity. Place identity strengthens one's sense of ownership of and responsibilities to the place, and further drives place-related behaviours, such as pro-environmental behaviours (Dresner et al., 2015; Dwyer et al., 2019; Halpenny, 2010; Meloni, Fornara, & Carrus, 2019). In tourism marketing and destination branding, place identity is also found to influence destination brand-building behaviours, such as active participation in tourism development (Chen & Dwyer, 2018), different types of word-of-mouth (Chen et al., 2014b, 2018), and intention to recommend (Chen et al., 2015). For example, the involvement of residents in strategic destination planning and policymaking is usually regarded as essential for sustainable tourism development, and the positive role of residents' place identity affects attitudes towards tourism (Chen, Wang, & Xu, 2017; Gu & Ryan, 2008; Wang & Chen, 2015).

Tourists' place identity regarding a destination can motivate similar pro-environmental behaviours as residents (Cheng, Wu, & Huang, 2013; Tonge, Ryan, Moore, & Beckley, 2015; Weaver, 2013). This is reasonably important considering tourists' often lack a sense of responsibility during travelling to unfamiliar destinations (Ettinger, Grabner-Kräuter, Okazaki, & Terlutter, 2020; Hibbert, Dickinson, Gössling, & Curtin, 2013b). Again, similar to the resident context, place identity is found to have a significant impact on revisiting intention, as well as intention to recommend (Tsai, 2016), implying the usefulness of place identity in encouraging consumer behaviours that benefit tourism destinations.

In addition to the place identity-behaviour relationship, in both resident and tourist contexts place identity is undoubtedly an important travel motive. Immigrants and their children (second-generation immigrants) often frequently travel to their places of origin, so as to retain or (re)establish virtual and physical connections, based upon their diasporic place identities, leading to a diaspora tourism market (Coles & Timothy, 2004; Hall & Williams, 2002; Huang, Ramshaw, & Norman, 2016).

First-generation immigrants are often overwhelmed with nostalgia enhanced by their place identities and are thus motivated to revisit their homelands (Basu, 2004). For second-generation immigrants, however, who may have never been to the places that their parents refer to as homelands, their place identities can have developed without physically visiting some those places. As discussed in Chapter 3, one's sense of place can be fostered purely based on narratives rather than physical engagement. For these second (and plus)-generation immigrants, their place identity associated with their homeland is more symbolic and ideological. Second-generation immigrants may learn about these so-called 'homelands' from their parents, relatives, family friends, and the media, thereby shaping their perceptions and constructing their sense of place purely in the imagination (Huang et al., 2016).

On the basis of diasporic place identity, several different outcomes may be expected from these second (and plus) generation immigrants visiting their homelands. First, a higher level of loyalty and place attachment may be created, due to less cultural and information gaps between these visitors and these destinations. Second, a significant discrepancy may occur between these visitors' pre-perception and perceived reality, leading to a serious cognitive dissonance. Third, the experience of diaspora tourists is affected by both their bonds to two or more places (this homeland and where they currently live as well as other locations), and their feelings are complicated by their confusion in balancing their multiple identities. This may be especially significant for younger individuals.

Box 5.4 Two generations of diaspora tourists in China

The Three Gorges Project (TGP), a major dam project resulting in hundreds of thousands of people being relocated throughout mainland China, lead to the existence of many migration villages in different provinces. Wushan 'migrant village in the city' is one of them.

In 2001, a village of around one thousand residents from Wushan, Chongqing was relocated to the city of Zhaoqing, Guangdong province. By 2019 a second generation of migrants villagers, who were born and raised in the new relocation city, had reached teenagerhood. The Wushan migration village had adapted itself to local culture, yet retained much of its original identity. Take the dialect as an example: older Wushan immigrants speak Chongqing dialect, while Zhaoqing local residents speak Cantonese. The older generation in this village have learned some simple Cantonese and their pronunciation can be easily picked up by locals. The younger generation who were born in Zhaoqing, on the other hand, do not speak Wushan dialect but speak fluent Cantonese.

The collision of two cultures cultivates quite a multiple and dynamic place identity in this village. The place identities of this village are reflected in three aspects: (1) culture, customs, and behaviours; (2) self-identification and categorisation; and (3) perceptions from outside the community.

In terms of culture and customs, the older generation in the migration village are more Wushan (place of origin) oriented than Zhaoqing (relocation city), although they embrace local culture and are willing to remain in line with locals. However, they feel obligated in reserving Wushan identity, but would prefer to be perceived as Zhaoqing locals.

Interestingly, the neighbourhood community close to the migrant village have gradually accepted these immigrants as locals, while other locals who do not know the background story simply assume they are not. Therefore, adult immigrants are constantly reminded that they are 'NOT Zhaoqing', often by a follow-up question, as previously illustrated, 'Where are you originally from?'

> When I am outside my neighbourhood talking to some local stranger, they would ask me again and again where I am from. Zhaoqing is not a satisfactory answer, and I have to say Wushan. I know that they have no ill intention, but I feel always reminded that I am not a 'real' Zhaoqing local. (An adult Wushan immigrant)

The younger or second-generation, specifically, are almost one hundred per cent Zhaoqing, with almost no mark from their parents' place of origin. Although often reluctant, they acknowledge their cultural and historical root of Wushan, and travel to Chongqing occasionally accompanying their parents to reconnect with their relatives, friends, and hometown area.

> I know and understand my roots in Wushan, although I have never been there. That town my parents used to live in is under water now. I am familiar with Chongqing cuisine. So every time I go to Chongqing with my family, I do not feel difficult with their local food. But I do feel

> bored since all my friends are here in Zhaoqing. (A teenage Wushan immigrant)

Older generation Wushan immigrants' Wushan identity drives them to revisit their homeland with their family, while their Zhaoqing identity drives their local citizenship behaviours. This is straightforward in their individual context. However, in a social context, how their place identities work and influence their behaviours is much more complex in terms of specific social scenarios.

As for the second-generation Wushan immigrants, their initial Wushan place identity was created from narratives and stories. This identity, however, is not powerful enough to motivate them to visit their homelands themselves. Their relationship with Wushan is still casual and undecided. They do not resent their Wushan identity, but they also do not see value in retaining it. With respect to diaspora trips to Wushan, they see it as purely as a 'family task', although this may be something that changes over time.

Case taken from unpublished fieldwork by Ning (Chris) Chen and Yong Rao in Zhaoqing, China.

Conclusions

This chapter has discussed the conceptualisation of place identity, and how it is connected to sense of place and place attachment in both theory and practice. Place identity strongly reflects the correspondence of self to place. As the chapter, and many readers, will note, as lives have become more mobile this also opens up the possibilities of multiple place identities and attachments. These are especially important issues for migrants but for tourists as well, as well as the ways in which migration and tourism are entwined. For example, in VFR travel, roots tourism and the overall desire to connect to place.

Issues of place identity are clearly of potential significance for destination branding and marketing and also connect with much deeper issues of tourism motivation. For example, our mobile lives mean that we often have multiple place identities and attachments. Tourism becomes one of the main ways by which those connections are maintained. Social media can help connectivity but over time it is really difficult to substitute a Skype or Zoom call for an actual face-to-face opportunity to meet, talk, share food, and the things one usually does with friends and family. The experience of COVID-19 shows some of the issues when people cannot do that. Therefore, the incapacity of many to travel to particular places because of the pandemic has had significant implications for place and personal identity, as well as feelings of a lack of connectivity, the longer term implications of which we are only now beginning to appreciate. However, the core message is that place and personal identities are entwined both in places of residence but also the places you travel too.

6 A holy mess? Sacred place and the religious sense of place

Introduction

Nearly all religions have explored the relationship between humans, nature, and God(s) (Wang et al., 2020a). Through a belief in a higher being(s), religion guides many people on not only how we behave but also what we do. Buddhism and Taoism, for example, emphasise that humans are part of nature and therefore should harmoniously coexist with the natural environment (Wang et al., 2020a). Christianity, and especially Franciscan Catholicism, stress that the resources surrounding humans were created and given by God and that God exists in all of nature (Wang et al., 2020a). Such examples highlight the connection between humans and religion in relation to the role of nature in facilitating this connection. As such, this chapter examines the role of religion in fostering sense of place and its related concepts of place attachment and place identity. The chapter starts by examining how religion, through the physical features of places and the socialisation that occurs in such spaces, contributes to a sense of place among pilgrims. Religion has also been shown to play an important role in place attachment (Mazumdar & Mazumdar, 2004). Recognising that the secular and post-secular can also contribute to sense of place (Hall, 2006a), the chapter progresses to examine spirituality and sense of place. Through travel, tourism objects (e.g., souvenirs), and religious practices at tourist sites, sense of place can be fostered. Thus, the last section of the chapter examines the 'holy mess' between tourism, religion, spirituality, and sense of place. The chapter concludes with a case study on Mecca and sense of place for Muslims.

The starting point for understanding the influence of religion on sense of place is related to physical features of nature such as rivers, lakes, and mountains having no intrinsic sense of place by themselves, but if bestowed with religious or spiritual meaning, these can foster sense of place, with some religions such as Buddhism and Hinduism, as well as many traditional indigenous belief systems, treating some physical features of places as living beings. Religion, therefore, has a profound influence on people's relationships to places and on place itself through physical features. Beyond the natural characteristics, sense of place can also emerge from human-made structures such as the design of cities, sacred spaces (e.g., temples and convents), neighbourhoods, and home (Mazumdar & Mazumdar, 2004). For the religious tourists, places with strong religious significance (e.g., Jerusalem) can help to establish self-identity (Mazumdar & Mazumdar, 2004). The development of self-identity is not restricted to making distinctions between oneself and others but also extends to objects and things, and the very spaces and places in which they are found (Proshansky et al., 1983). Place identity, therefore, is a sub-component of the self-identity based on cognitions about the physical world in which the individual lives (Proshansky et al., 1983). A person's religious self can be linked to places and spaces significant to his/her religion (Mazumdar & Mazumdar, 2004). Given that sense of place can give expression to emotions through the link to affective qualities of places, intensifying the religious qualities of places can, arguably, contribute to a sense of place that encourages destinations, regions and countries to recognise, protect and enhance those qualities that contribute to place character.

Tourism, religion, and sense of place are intertwined in a way that highlights many of the essential aspects of being human. Yet, within the tourism literature, concepts such as religion, religiosity, and spirituality have been described as a 'holy mess' (Poria, Butler, & Airey, 2003). For example, there is no agreement on the individual motives influencing religious site visitation. While some argue that these may relate to faith, spirituality and piety (Buzinde et al., 2014), others suggest they are related more to curiosity, cultural and heritage motives, and discovery rather than religious motives (Amaro et al., 2018). This tension can be related to the traditional focus of tourism studies on pilgrimage, which was exclusively religion-based, to more recent recognitions of the post-secular pilgrimage and more general issues of spirituality, faith, and religion being examined in tourism practices (Hall, 2006a). The overwhelming conclusion is that these issues are intertwined and multifaceted.

We live in post-secular times, in times when faith, religion, and spirituality are reclaiming secularised spaces of modernity (Nilsson & Tesfahuney, 2016). Tourism mobilities, specifically pilgrimage tourism, have been central to what has been described as the resurgence of religion/spirituality in contemporary Western societies (Nilsson & Tesfahuney, 2016). Apparently the marginalisation of religion in modernity creates a spiritual vacuum in modern life, which the post secular is trying to (re)fill with meaning. Thus, the post-secular condition is seen as a reaction to the secular, as the secular was a reaction to the religious past (Sigurdson, 2010). New Age religions and belief systems are, therefore, flexible and recognise that there are many paths to God(s) as there are people, and are mindful of individual choices and freedom of faith (Nilsson & Tesfahuney, 2016). Individuals conceive of and relate to post-secular religiosity as a source of personal, social, and moral commitment that offers an ethical orientation towards self, others, and creation at large (Nilsson & Tesfahuney, 2016). Thus, sense of place is also related to post-secular pilgrimage tourism (Nilsson & Tesfahuney, 2016).

Religion and sense of place

The literature on religion and sense of place is miniscule in comparison to that on the natural environment and sense of place. Nevertheless, the two distinct research strands overlap when religion is considered. Previous studies suggest that active engagement with the landscape is inherent to and inextricable from the practice of religion (Allison, 2019). Through religion people express love for nature, often called topophilia or love of a landscape (Tuan, 1972). As outlined earlier, the basis of many religions is the relationship between humanity and God and that of humans and nature. For example, the pan-Himalayan countries (e.g., Nepal, Tibet, and Bhutan), which have a strong influence of Buddhism, Hinduism, and Taoism, tend to see the landscape as lively and animated, full of protector deities, and local spirits that are perceived to inhabit various features of the landscape (Allison, 2019). For example, local spiritual beliefs related to the landscape are prized as central to the cultural identity of the Kingdom in Bhutan (Allison, 2019). As a consequence, rivers, mountains, lakes, and ponds are given special symbolic meaning as protectors of the Kingdom. Religious places are more than just mere beliefs and affect, they are also location for behaviours, including worship, offerings, giving alms, and bathing (Ruback, Pandey, & Kohli, 2008). These authors use the example of the *Sangam* (the confluence of the Ganges and Yamuna rivers) in Allahabad, India, which is a popular pilgrimage site known as *Prayag*, as an example to illustrate how sacred places are endowed with different material, social, and symbolic aspects, and thus contribute to identity formation.

Religion endows places (natural landscapes and man-made) with symbolic meanings, which not only help to differentiate them from ordinary spaces, but through their geography, design, or architectural aesthetics can have the capacity to foster attachment, devotion, spirituality, and a 'worldview' (Mazumdar & Mazumdar, 2004). From the perspective of sacred spaces, Mazumdar and Mazumdar (1993) examined the interconnectedness of religion, identity, and place attachment. Their research suggests that place attachment and identity emerge from the ways religious rituals connect people to places and how religious settings become places for socialisation that connect people to religion. While Proshansky (1978) tends to describe self-identity as a singular concept, Mazumdar and Mazumdar (1993) argue that self-identity is composed of the spiritual self and a religious self. The spiritual self and identity are intimately connected to the religious self and identity (Mazumdar & Mazumdar, 1993). In their work, these authors recognise that place attachment can emerge from religious spaces within the home (e.g., prayer rooms, religious statues, and altars) and through attachment to religious architecture (e.g., temples and churches). Other facets of place attachment can be related to religious cities and locations (e.g., Canterbury, Jerusalem, Vrindavan, and Mecca, and/or their natural surroundings). The example from Box 6.1 illustrates some of these different forms of attachment by examining the significance of the lake, *Ganga Taloa,* in Mauritius for the celebration of the *Maha Shivaratree* festival.

In a later study, Mazumdar and Mazumdar (2004) propose a multifaceted model of religion and place attachment, whereby the place (physical characteristics and qualities), religious socialisation in the place, and individual or collective attachment can explain how religion affects place attachment. According to this model, a person's connection to place is not solely reliant on the qualities of place but also on the learning process associated with religious socialisation. This

Box 6.1 Ganga Taloa, pilgrimage and place attachment

The annual pilgrimage of Maha Shivaratri is a century old tradition and is the largest religious festival on the island of Mauritius. The festival, which is associated with Hinduism, consists of a pilgrimage to a sacred lake, *Ganga Taloa* (Ganges pond) or Grand Bassin (great lake), to pray and make religious offerings to the lake and Lord Shiva. The lake was discovered in 1897 after its location was revealed to a priest in a dream. According to a myth, it was formed when Shiva spilled some drops of the Ganges, which he kept in his head. In 1972, water brought from the Ganges was poured into the lake, symbolically reaffirming its connection to Hinduism's most sacred river (Xygalatas et al., 2016, 2018). The lake, therefore, symbolises the river Ganges in India. The pilgrimage draws hundreds of thousands of pilgrims of all Hindu denominations every year, as well as people of other faiths. Although not barred from visiting the place, non-Hindus on the island do not necessarily feel a sense of belonging to the site (Boswell, 2005). In recent years, an enormous statue of Lord Shiva, 108 feet tall, has been erected near the lake, welcoming pilgrims to their destination. During the pilgrimage, devotees walk for many miles, often for the entire duration of the trip, which can take several days. When they reach the lake, they make offerings to the god and collect sacred water to bring back with them. The water is transported in large, temple-shaped structures called *kanwar*s and is eventually poured over the Shiva Lingam (a symbolic representation of Shiva) at the pilgrims' homes and temples (Xygalatas et al., 2018).

This festival epitomises several facets of people's attachment to sacred spaces. In particular, Mazumdar and Mazumdar's (1993) attachment to landscapes, architecture, and homes are evident from this festival. First, the lake itself and the surrounding landscape which consists of a forest, monkeys, and hill are sacralised and revered. The water from the lake is used for ritual purification and offered to the Gods. Pilgrims also make offerings of coconut, incense wood, flowers, and other fruits to the lake. This attracts the monkeys who feed on these offerings. In Hinduism, monkeys represent Lord Hanuman (the monkey lord) and are venerated as protectors of humanity. Accordingly, the monkeys at the lake have a special status with Mauritian Hindu folklore. These exemplify how natural landscapes associated with sacred places can become a source of place attachment. Second, the architecture of the various temples located at the lake is also a source of place attachment. These temples are dedicated to Lord Shiva, Lord Hanuman, Goddess Durga, Goddess Ganga, and Lord Ganesh, among others, and therefore each has a distinct architectural style to commensurate with Hindu beliefs about each God. The sanctity of the place is conveyed through collective prayers and rituals during the festival and the rest of the year. Third, the water from the lake is brought home before it is poured onto the Shiva Lingam. Home becomes sacred not only because of the water being kept at home but also because gods, spirits, and ancestors are worshipped at home and objects such as prayer rooms, pictures of Gods and Goddesses on the walls, statues, and sacred icons are all part of practising Hinduism and celebrating Hindu festivals. Therefore, the festival not only serves to connect pilgrims to God but also to home and the natural environment. In this way, place attachment develops for Ganga Taloa.

religious learning through priests, educators, and peers can play an important role in identity formation through teaching prayers, performing religious rituals, telling religious stories, and through personal experience with religious places. This study places greater recognition on the evolving identities and roles of pilgrims, their experiences and interactions with pilgrimage sites, in shaping religious and self-identity of individuals. While the majority of studies examine sense of place in relation to religious sites (physical characteristics), material nature can also play a role in the religious experience with objects, e.g., souvenirs, holy water and pieces of the sacred ground, allowing for the performance of rituals and expressions of beliefs (Terzidou, 2020). Material objects are a significant part of Christianity, Buddhism, and Hinduism, where ritual objects can have a profound religious effect. Through use of religious objects, self-identity is affirmed that differentiates its owners from others and protects the person from unforeseen events (Terzidou, 2020), and therefore can contribute to sense of place as much as the physical characteristics of place.

Religious belief systems may transcend attachment to any particular place (Counted & Watts, 2017), which has led researchers to also examine the relationship between the concept of religiosity and sense of place. Religiosity can be conceptualised as the degree of one's belief in God and a commitment to follow the principles set by God (Bhuian et al., 2018). Given that religious significance and the relative devotion of pilgrims can to some extent be used as surrogates of their identities (Belhassen et al., 2008; Della Dora, 2012), the relationship between place identity, place dependence, attachment, and religiosity has been investigated in a fragmented way, with no consensus on whether religiosity affects all of these concepts and to what extent. Nevertheless, religiosity has intrinsic and extrinsic dimensions. The former refers to the religious orientation of a person where he/she engages in religion and its practices for achievement of internal goals related to the religious traditions and practices themselves. The latter refers to the religious orientation of people for achieving their own personal goals (Allport, 1966). Individuals with intrinsic religiosity tend to view religion as the ultimate end in itself while those with extrinsic religiosity tend to view religion as an instrument to achieve certain goals (Allport, 1966). Therefore, there may be differences in the way these two types of religiosity influence place attachment and place identity. For example, whether an individual values physical characteristics of place and therefore engages in protective behaviours about that place may be related more to intrinsic rather than extrinsic religiosity. Religiosity has also been shown to have a strong effect on pro-environmental behaviours (Wang et al., 2020).

From the lens of disaster recovery (see Chapter 7 for more details) Dueck and Byron (2011) found that the religious importance of places and spiritual connections with land supported recovery pathways for individuals and communities. Their findings convey the importance of religiosity as shaping a collective rather than an individual approach towards valuing the landscape in shaping disaster recovery. The authors emphasise the importance of collectiveness in non-Western communities that is often overlooked and understated in post-disaster recovery (Dueck & Byron, 2011).

The intersection between beliefs and place is often referred to as 'geopiety' (Tuan, 1977) and, therefore, can be described as the cognitive and affective attachment to a sacred place that is based on faith and values (Belhassen et al., 2008; Kim et al., 2020a). There is a recognition in this literature that the discourse on place can be about meanings, experiences, or the objective environment, but these are not independent of each other (Belhassen et al., 2008). Therefore, sense of place emanating from the natural environment can contribute as much as to the meaning of the religious site itself as well as to the authenticity of the experience. In this regard, the experienced authenticity at holy sites can be related to pilgrims' belief, activities, and places, referred to as the concept of *theoplacity* (Belhassen et al., 2008). This concept integrates ideological and spatial dimensions in evaluating the authenticity of religious sites. More recently, the concept of spiritual attachment has been used to further elaborate on the influence of religion on sense of place. Spiritual attachment reflects the normative ways in which people evaluate and experience their spirituality in terms of developing proximity to God, turning to God as a response to loss, and perceiving God as a safe heaven and a secure base (Counted, 2016, 2019). In this way, the physical and social characteristics linked to religious places can foster spiritual attachment. For example, many pilgrims travel to the French town of Lourdes each year to visit the Grotto of Massabielle (cave), where in 1858 the Virgin Mary appeared to a local woman. Since then, many alleged miraculous healings have been documented at this site. The town itself is surrounded by three summits reaching up to 1,000 m, and therefore not only the religious meaning of the site but also its natural landscape contributes to attachment to the town as whole. Also, given that those undertaking this pilgrimage are often undertaking it to cope with loss or to seek a cure, it can be argued that their spiritual attachment is driven by a perception of God as a saviour and place as a safe heaven.

Spirituality and sense of place

While the previous section examined sense of place in relation to religion, in post-secular times there has been a greater emphasis on spirituality rather than religion. Also, pilgrimages need not necessarily be religious in nature to be endowed with meaning (Di Giovine & Choe, 2019; Hall, 2006a). Geographic places and divine entities can be perceived as objects of attachment from a spiritual lens (Counted & Zock, 2019). Individuals look for spiritual experiences beyond and outside of the conventional religious symbolisms associated with places of pilgrimage (Collins-Kreiner, 2010). The spiritual encounter (felt or imagined), the feeling or sense of revelation, can be tied to the site or place without any religious connotations (Hall, 2006a; Nilsson & Tesfahuney, 2016). Traditional definitions of spirituality have become fragmented as they become linked to secular overtones (Sharpley, 2009), which implies a drive towards being located within religion but also distancing from religion (Stausberg, 2014). The spiritual realm assumes that the environment has symbolic and affective properties and is an important vessel for cultivating identity and values (Ruback et al., 2008). Group experiences and activities occurring in spiritual places (e.g., Stonehenge) also acquire social meaning (Ruback et al., 2008). Before the advent of Christianity, many pagan religions were centred on the recognition of the divine in nature and a strong awareness of the natural world and the power of the divine in the ongoing cycle of life and death (BBC, 2002). Research on traditional societies in Africa and South America confirms that local people and spiritual cosmology are mutually constitutive, where local people's belief and sacrifice for the spiritual realm perpetuates its existence, while the spiritual cosmology shapes local identities and creates a unique sense of place and interaction with the environment (Golden, 2014). One form of this spirituality is how communities view their relationship with the land. For Māori, land not only gives but also shapes identity (see Chapter 2). Local people can treat land as both productive in its utility and also as a space of spiritual communion and moral structure (Golden, 2014). The spirit world, therefore, requires local people's belief in them to them to perpetuate their existence. In return, the spiritual world's benevolence furthers humanity's success, and more broadly, structures local identities as intimately connected to their environment and the spiritual world it houses (Golden, 2014).

As described above, human interactions with the environment is a source of spirituality and this relationship has received some attention in the literature (Golden, 2014). Others offer insights into how spirituality affects sense of place (Counted, 2019; Counted & Watts, 2017). On one hand, the expression of attachment to the divine in certain places among different groups has been documented (Counted & Zock, 2019), thereby suggesting that sense of place fosters spirituality. On the other, the search for individual spiritual experiences can change the meaning of places by shifting place identity. Places traditionally associated with religion can gradually shift from a place whose meaning is tied primarily to a collective religious identity into one that is fleeting, porous, and personal or individualised (Nilsson & Tesfahuney, 2016). Tourists walking the *Camino de Santiago de Compostella* without any religious obligations still do so in the hope that they will find some kind of meaning from the experience (Norman & Pokorny, 2017). These self-described pilgrims can also make pilgrimages to non-religious sites such as memorials and other sites of loss, heritage sites, parks, and landforms, and to other places that hold particular importance to them as an individual (Di Giovine & Choe, 2019). Repeat walking pilgrims, however, are primarily motivated by spirituality (Amaro et al., 2018). These pilgrims seem to be seeking partly new experiences related to nature (Amaro et al., 2018; Kim et al., 2020a). Such influences can diminish the associated religious meanings and symbolisms and intensify the touristic meanings of place. However, touristification can also spur a resurgence of religion and of post-secular pilgrimage in a quest to re-establish place identity and meaning for both believers and non-believers. The experience of a sense of place is, arguably, different from the original pilgrims that visited such places and therefore, place identity is shifting constantly with post-secular place/landscape pilgrimage. There seems, however, to be some similarities between the traditional religious and current post-secular pilgrimages in that, in both, elements of existentialism, fulfilment, and forsaking material possessions are evident (Nilsson & Tesfahuney, 2016).

More recently, the term place spirituality has emerged (Counted & Watts, 2017) and can be conceptualised as a transactional process and an organised attachment strategy between an individual and his or her objects of attachment (Counted & Zock, 2019). These studies emphasise the role of spirituality in, for example, managing stressors associated with a place, thus conceptualising places as distal dynamic contexts of social interaction that contribute to spiritual interconnectedness (Counted, 2019). Related to spiritual attachment is the concept of mindfulness that would allow individuals to attach to places and objects. Mindfulness practices are a means of dealing with negative life events, and therefore would resolve problems related to depression and anxiety (Choe & O'Regan, 2020). In this way, there is a need to understand the nexus of people and place spirituality in relation to cases where place attachment might be disrupted by disasters, war or conflict, and terror attacks. However, the notion of place spirituality has yet to be embraced in the mainstream place attachment literature. Next, the complex relationship between tourism, religion, and sense of place is examined.

Tourism, religion, and sense of place

Tourism and religion have always been intricately linked. It is a common motivation for travel and remains one of the world oldest forms of mobility (Collins-Kreiner, 2010). A diverse range of religions has been studied in tourism including Christianity, Islam, Hinduism, Buddhism, and Judaism (Collins-Kreiner, 2020). However, this field of study has been approached from diverse disciplines ranging from anthropology to economics, psychology, and management (Kim et al., 2020a; Timothy & Olsen, 2006). Religious tourism encompasses different activities such as pilgrimage, faith-based events and festivals, and devotional sites, among others. While early studies of these phenomena have been situated within religious, heritage, and cultural tourism discourses, more recently the topic has been linked to eco-tourism, sustainable tourism, and slow tourism. There is a recognition of the blurred boundaries between religion and tourism and among different tourism segments (Collins-Kreiner, 2010; Di Giovine & Choe, 2019). Several reviews of this literature (Collins-Kreiner, 2020; Durán-Sánchez et al., 2018; Kim et al., 2020a) provide no consensus as to what religious tourism is all about. Religious tourism can be understood, for example, from the lens of visitor identities and the destinations they go to and therefore can include examinations of visitors' religiosity in relation to their religious affiliations and particular places they visit (religious, holy, and sacred sites) (Kim et al., 2020a). Religious tourism can also be understood from a cultural approach examining pilgrimage and tourism as modern practices in a post-modern world (Durán-Sánchez et al., 2018). Another approach is to consider tourism as a contemporary spiritual journey (Sharpley, 2009). This recognises that religious tourism has evolved beyond pilgrimage to encompass the meaningfulness of a destination (Kim et al., 2020a). This implies a cross over with spiritual tourism and mindfulness (Choe & O'Regan, 2020). The term spiritual tourism has no universally accepted definition, it arguably comprises of travel by participants who are seeking a range of desirable outcomes, religious or non-religious, where the underlying motive guiding the travel is the yearning for challenge, visceral, intellectual, transcendent, and transformative experiences (Cheer et al., 2017).

Irrespective of the approach used by scholars, religious or spiritual tourism epitomises an emotion laden travel, encountering places, landscapes, and material objects that can activate the five senses in an attempt to move into new realms that entail a connection with God, the spiritual, non-human, and also the unknown. For example, through bodily contact with religious objects, sacredness can be experienced and collected through feelings of awe and serenity (Terzidou, 2020). Also, as discussed earlier landscapes can become vessels for both religious and spiritual experiences. Della Dora (2012) highlights that the wilderness and the holy share two main similarities. First they both have affective attributes which can cause wonder, awe, and surprise as well as generate inner clam and serenity. Second, they both evoke the separation from the ordinary against which they are defined (Della Dora, 2012). It is perhaps therefore no surprise that some of the early travel to national parks in the United States was inspired by literature and images that spoke of the early park experience in spiritual terms (Hall & Frost, 2009). In a study of pilgrims to Mount Athos and Meteora in Greece, Della Dora (2012) highlights the importance of the landscape to experiences of pilgrims at both sites. While some tourists visit spiritual landscapes which may or may not be associated with religion (e.g., the Santiago de Compostela pilgrimage), others visit sites which have cultural connections to ancient civilisations (e.g., Macchu Picchu), or may be the original homeland of First Nations (e.g., Native American Indian or Aboriginal landscapes). Pilgrimage tourism spaces, therefore, become open-ended spaces where place identity and meaning are being continuously reworked (Nilsson & Tesfahuney, 2016). In tourism studies on spiritual experiences, though sense of place is not often examined explicitly, there is an implicit assumption that spiritual tourism fosters sense of place. For example, Sharpley and Jepson (2011) found that rural tourism can be a vessel for spirituality and from this experience tourists obtain a sense of place.

It is not surprising that religious tourism has also been discussed in terms of spirituality rather than religious motivation and actions in recent years (Collins-Kreiner, 2020). Different types of tourists are motivated to travel to religious or spiritual destinations by a quest for spiritual well-being, enlightenment, knowledge, and social relationships (Collins-Kreiner, 2020). Spiritual tourism has been popularised by Hollywood movies such as *Eat Pray Love* and tourism imaginaries about particular people and places. Destinations such as Ubud in Bali have become a sacred landscape and act as a 'faithscape', which embodies not only tangible geographical features but also their intangible spiritual elements

(Singh, 2006). This confirms that sense of place can also emerge from the feelings of well-being that can be enhanced through contact with natural landscapes. However, spiritual tourism has tended to be portrayed as a reflexive well-being intervention driven by the sense that some aspects of everyday life need improving (Norman & Pokorny, 2017). Spiritual pilgrims tend to, however, distance themselves from the traditional religiosity inherent to pilgrimage (Cheer et al., 2017). Emphasis is placed in spiritual travel on flexible and negotiable individual and social identities (Olsen & Wilkinson, 2016). This implies shifting and evolving identities of both the spiritual and religious tourists, recognising that there are different pathways for experiencing spirituality through travel and encounter with people and landscapes. The motivations that guide such travels are subjective and dependent on imaginary boundaries (Della Dora, 2012). Spiritual tourists through exploring pilgrimage sites and engaging in reflective activities pursue their own sense of meaning, purpose, and identity (Norman & Pokorny, 2017). This implies that place identity and place attachment are the result of reflective practices.

Secular pilgrimage

The concept of secular pilgrimages has appeared to emphasise journeys to non-religious sacred sites that may have spiritual significance. These emerge to provide a more granular and holistic understanding of the relationship between tourism, religion, and spirituality (Hall, 2006a; Kirillova, 2019). According to some authors, we live in post-secular times in which faith, religion, spirituality, and mindfulness are reclaiming secularised spaces of modernity (Nilsson & Tesfahuney, 2016). There is an overshadowing of sacred places and traditional pilgrimage destinations by spiritual journeys (Olsen & Timothy, 2006). This implies that the blurred boundary between official sacred space and secular space is not simply relegated to scholars, but is implicated in complex discourses and practices of travellers, hosts, and multiple stakeholders (Di Giovine & Choe, 2019). Post-secularism can be seen as a new religiosity that individualises faith and spirituality (Nilsson & Tesfahuney, 2016). Post-secular religiosity focuses on individual, private, faith, and practice (Ziebertz & Riegel, 2010). It has more to do with questions of existence, meaning, fulfilment, and well-being, rather than faith sanctioned by the expectations and obligations of institutionalised religion (Nilsson & Tesfahuney, 2016; Ziebertz & Riegel, 2010). It is clear that spiritual tourism contributes to debates on the nature and meaning of human existence and that existential thought and faith are not mutually exclusive (Kirillova, 2019). There are several examples of tourism products such as wellness retreats that offer 'transformative' experiences meant to challenge the tourist worldview, help solving personal dilemmas, or simply to (re)discover oneself (Kirillova & Lehto, 2015). For example, Prayag et al. (2015) elaborate on the experiences of tourists consuming ayahuasca in so-called spiritual retreats and the transformative experiences that follow. They highlight that such experiences are very much dependent on expectations, the shaman performing the rituals and the motives visiting the retreat (Prayag et al., 2016). Yet, little research focus on existentialism in the areas of tourism and religion, which are typically centred on pilgrimage experiences (Kirillova, 2019) and its connectivity to place and being.

Box 6.2 Mecca, sense of place and the journey home

Islam is a religious faith founded in a belief in *Allah* (God) as the sole deity and in *Muhammad* as his final prophet. The religion is founded on five pillars: confession of faith (*shahada*), implying that there is no god but Allah who is the only supreme being and Muhammad as the final messenger; *Sallat* which involves performing five prayers a day; *Zakat* which is about donating annual wealth to the poor; fasting during Ramadan; and making a pilgrimage to Mecca (Ghaderi et al., 2020). Of all the major pilgrimages in the world, the hajj is not only the largest but the only one authorised in scripture as one of the five pillars of Islamic faith (Di Giovine & Choe, 2019). In Mecca, a Muslim finds places significant in the life and actions of the Prophet and Islam. Located there is the *Kaaba*, a 'large cubic structure… draped in gold-bordered black silk cloth… the centre point of the religious world for Muslims' (Bowen, 2002, p. 228). The *Quran* says, nothing, however, about the remote origins of a holy place at Mecca, it speaks only of the era of Abraham and of Ishmael there, and of the providential construction of the *Kaaba*, this 'sacred house' (5:100), this 'ancient house' (22:29), that sat in the midst of the town (Peters, 1996). The place itself, Mecca, helps to establish the self-identity of believers. As such, Islam is like many other faiths, in that a person's religious self can be linked to places and spaces significant to his/her religion (Mazumdar & Mazumdar, 2004). However, while pilgrimage is religious for some, it can also be secular for others (Delaney, 1990).

For Muslims, the *hajj* pilgrimage to Mecca commemorates Prophet Muhammad's pilgrimage to the city. It is a religious duty requiring all Muslims to undertake this journey at least once in their life. In Mecca, pilgrims undertake several prescribed ritual acts (Bowen, 2002). The Muslim's hajj is more than just a visit to a holy place or shrine as will be illustrated in this case. It is a place for reaffirming religious identity, building collective attachment, and form attachment with material objects. Pilgrimage as a religious requirement has important implications for place attachment. It emphasises the importance of place associated with the Prophet, and therefore the sacredness of the space itself. All believers experience its physical qualities and identification with place becomes an integral part of Muslim identity (Mazumdar & Mazumdar, 2004). The hajj is a unifying experience that integrates the pilgrim's present life with both the past and the present. They are thought to be vouchsafed a glimpse of the other world, which they imagine to be their original, true, and final home (Delaney, 1990). This implies a strong connection with the unknown that reinforces self-identity and reinforces spiritual meaning of the pilgrimage. The experience also is a learning process associated with religious socialisation. Believers learn stories about the Prophet and his time in Mecca at home and in mosques and are able to relate these stories to sites of significance during their pilgrimage in Mecca. The religious self and self-identity become one during the experience. However, the journey starts with the preparation for the *hajj,* which can involve a series of stages encompassing material, spiritual, and informational dimensions (Caidi, 2019). This demarcates the trip from the ordinary religiosity of Muslims by emphasising the once in a life time aspect, which requires more planning than other forms of pilgrimage. Pilgrims' information practices are varied, and transcend both individual (cognitive, affective) as well as social processes (through shared imaginaries, and a translocal network of people and resources) (Caidi, 2019). *Hajjis* will seek information and blessings from those who have been from their local mosque.

The pilgrimage through the performance of rituals by thousands of others attests to Mazumdar and Mazumdar's (1993) concept of collective attachment that fosters place attachment. For many Muslims to make the hajj is to touch the foundation of faith and to drink from the wellspring that sustains and gives it meaning. The spiritual meaning is symbolised concretely by pilgrims touching or kissing the black stone nestled in the *Kaaba* and drinking from the well of Zamzam that sustained Hagar and Ismail in the desert (Delaney, 1990). Pilgrims wear special clothing and are required to collectively circumambulate the *Kaaba* seven times, run back and forth along a street in Mecca, stand together in the plains of Arafat, and stone three pillars at Mina (Bowen, 2002). These actions performed sequentially engage the pilgrim in the sacred geography and history of the place (Mazumdar & Mazumdar, 2004) and, given that these are performed by thousands of others, reinforce collective attachment to the site through this unique Islamic rite. In this way, religious places such as Mecca become the focus of intense collective place attachment. Through collective rituals, ties to place, community, and religion are formed and transformed (Mazumdar & Mazumdar, 2004). The place is experienced through the senses: seeing sacred sights (the Kaaba), hearing sacred sounds (prayer callings), touching sacred artefacts (the black stone), eating special food, and smelling specific scents (incense) (Mazumdar & Mazumdar, 2004; Ruback et al., 2008).

While the onsite experience is purportedly transformative, the journey home is supposed to be socially transformative. The *hajjis'* status as evidenced by the level of respect they have in society is usually higher than when they left. A Muslim returning from *hajj* is seen as a dispenser of *berkat* (blessings) and therefore, family and friends come to visit for several weeks on return. As argued earlier, pilgrimage can affect identity. On their return home from Mecca, the place becomes part of their identity, as pilgrims use the honorific *hajji* (for males) and *hajjiya* (for females) (Mazumdar & Mazumdar, 2004). Upon their return, the *hajjis* await visitors and usually distribute the myrrh, frankincense, dates, baubles, bangles, and beads they have brought back with them. These gifts are of little monetary value but are highlighted as symbolic as they allow the visitors to experience a little bit of Mecca (Delaney, 1990). The *hajj* is therefore a clear rite of passage – a socially approved ritual intended to manage status change (Di Giovine & Choe, 2019). The return home also signifies a deeper connection to Allah and through prayer rituals, which include turning towards Mecca through prayers five times a day and in death, cement their connectedness to the sacred space that they have experienced (Mazumdar & Mazumdar, 1993).

Conclusion

Tourism, religion, and spirituality are linked in ways that make it difficult to distinguish between the secular tourist and pilgrim. Traditional definitions of pilgrimage, spirituality, and religious tourism are fragmented and imbued with secular tones (Sharpley, 2009). Nevertheless, it is clear that whether it is religious or spiritual tourism, the experiences can be emotional and transformative, thus fostering senses of place and well-being. While the broader literature on religion and sense of place highlights the importance of natural landscapes, the role of pilgrimage sites and the socialisation process that occurs with religion in the formation of place identity and place attachment, it is apparent that these outcomes do not materialise for every individual and across religious sites. What creates place attachment for one individual may contribute to place aversion in another. The emergence of new concepts such as place spirituality and spiritual attachment also muddies understanding of how religion and spirituality contributes to sense of place. The chapter also highlights the role of religiosity (intrinsic and extrinsic) in guiding behaviours. Tourists with intrinsic religiosity, for example, are more likely to follow their religion's doctrine, rules, and beliefs and voluntarily perform pro-environmental behaviours. These tourists also have a high sense of connectedness to nature. However, extrinsic religiosity may not always have an effect on pro-environmental behaviours (Wang et al., 2020). Where there is a worldview, from a religious or spiritual perspective, in which there is a symbiotic relationship between human, God/Divine, and the natural environment, this should in theory engender tourism practices that are more environmentally friendly and sustainable. These tourism practices while contributing to sense of place should also foster well-being and reflection on the tourist activity as a practice in itself. However, the reality of such behaviours and practices remain to be seen.

7 You don't know what you've got until it's gone! The dynamic nature of sense of place

Introduction

This chapter focuses on the changing and evolving nature of places by examining how sense of place is affected by unexpected or sudden change, e.g., disasters; continuous change, e.g., migration; and incremental change, e.g., environmental change. While recognising that changes at these three levels are often inter-related, the chapter first outlines the role of sense of place in facilitating or impeding how communities prepare for, adapt to, respond, and recover from disasters. Disasters are often not only sudden occurrences with limited time for planning by individuals and communities (Bonaiuto, Alves, De Dominicis, & Petruccelli, 2016) but can also change drastically the landscape and meaning of places (Nicolosi & Corbett, 2018). Therefore, by examining the different stages of a disaster (preparedness, mitigation, response, and recovery) it is possible to capture the physical changes to place as well as the socially constructed meanings of place. In this way, the opportunities and tensions that arise from disasters in relation to different facets of sense of place, e.g., attachment, identity, and rootedness, can be understood. For the tourism industry, disasters can change travellers' perceptions of destinations in regard to attractiveness, hosting capacity, access to tourism infrastructure, and quality of the tourism experience. In this way, disasters can alter the sense of place for tourists, affecting levels of attachment and identification with place. The case study of tourists' evolving sense of place based on their experiences of the servicescape emerging from the September 2010 and February 2011 Canterbury Earthquakes in New Zealand highlights some of these issues.

People can become placeless and displaced not only because of disasters but also due to climate change, war, forced migration, and economic difficulty (McKinzie, 2019). The second part of this chapter delves into the complex relationship between migration and sense of place. Whether migration is perceived as legal (e.g., immigrants) or not (e.g., asylum seekers), migrants can change the meaning of a place (Myat Thu, 2020) and be changed by it (Husa & Morse, 2020). From this perspective, sense of place can be both a positive and negative factor contributing to migrants' adaptation to their new environment. By examining different types of migration (e.g., rural to urban and vice-versa, and international migration) and sense of place, this chapter highlights the morphing nature of place identity and place dependence in relation to both the 'new' (settled place) and 'old' place (place of origin). From a tourism perspective, the tourism flows from the 'new' to the 'old' place through the lens of diaspora tourism or visiting friends and relatives (VFR) and the ensuing implications for tourist behaviour and the tourism industry are highlighted.

Both migration and disasters can alter the sense of place but changes to the natural environment can precipitate both disasters and migration. The impacts of environmental change on sense of place can be sudden (e.g., fires, storms, and flood), continuous, and incremental (e.g., average temperature increases as a result of anthropogenic climate change). The last section of this chapter examines how sense of place is related to environmental change by considering, specifically, the relationship between sense of place and climate change. In studies on environmental change, place attachment has been shown to act both as a barrier to transformative adaptation and motivation for action (Jeffers, 2019). The consequences of environmental change not only have implications on residents' sense of place but also that of tourists. Thus, changing landscapes, climate and weather patterns, and disappearing native species can affect destination attractiveness and sustainability. Next, the dynamic nature of sense of place in relation to different scales of change is discussed.

The dynamic nature of sense of place

Places are constantly changing, evolving, and stabilising, which leads to people having different perceptions of and attachment to places. This dynamic nature of place and its corresponding effect on sense of place and place attachment has already been considered in previous chapters (see Chapters 2 and 3). However, as a dynamic concept sense of place implies an ever changing relationship between place and the environment. Although sense of place is a concept that usually has positive connotations, the inherent dynamism that embodies the concept is neither always positive nor confers only benefits to individuals and communities. This is because of the shifting meaning and physical characteristics of places, which can not only signal progress, achievement, and stability but may also signify decay, destruction, and instability due to disasters, migration, and economic and environmental change. People can distance themselves from a place as a result of negative experiences and memories due to, for example, civil conflict or forced migration (Shamai, 2018) and disasters (Chamlee-Wright & Storr, 2009; Cox & Perry, 2011). In this way, a negative sense of place emerges. Experiences of displacement can also lead to feelings of helplessness because people feel that they have been stripped of their agency (McKinzie, 2019). As a result, sense of place becomes fuzzy and detachment from the place can ensue. While there is dynamism inherent in a place reinventing itself following a major negative event, whether or not a sense of place retains its original meaning hinges upon how place-related changes are portrayed and managed. For example, people's routines and customary ways of thinking are disrupted when disasters strike (Hall et al., 2016). Faced with uncertainty and the threat of the collapse of the meaning of place itself as well as its relationship to self, people are inclined to examine place–environment relationships for sense-making (Stein, 2004). Sense-making is essential for survival post-disaster (Stein, 2004) and for place to retain its identity and meaning for individuals and communities requires that the disaster recovery process is imagined and managed in a way that is inclusive of community expectations (Gurney et al., 2017). However, as Jeffers (2019) points out, sense of place as a boundary concept in such circumstances revolves around decision-making processes that are centred on communities, while this may facilitate the evaluation of new approaches to hazard management and climate change adaptation, it also poses significant challenges to change place (Jeffers, 2019). People can resist changes that have potentially the ability to modify sense of place both in relation to physical characteristics and social relationships in a place.

Sense of place and sudden changes as a result of disaster

Sense of place has a complex relationship with different stages of a disaster. This relationship certainly requires more research (Bonaiuto et al., 2016; DeYoung & Peters, 2016). Nonetheless, one common thread that emerges from existing studies that cover a wide range of disasters, including earthquakes, floods, and oil spill, is that disasters can alter sense of place for individuals and communities drastically. Through concepts of place attachment, place identity, place dependence, sense of place, and belongingness, existing research shows that place-based relationships influence disaster preparedness, recovery, and resilience (Bonaiuto et al., 2016). Sense of place can enhance community preparedness for hazard events (Mishra et al., 2010). Research on this relationship shows that self-efficacy of individuals and their sense of community can have a positive influence on disaster preparedness (DeYoung & Peters, 2016). A distinction should be made between perceived and actual preparedness (Kirschenbaum, 2002). In this, the notion of risk is also important, given that risk perceptions affect both perceived and actual preparedness (Paton et al., 2000). In fact individuals and communities may have different levels of awareness of hazards and hazard anxiety, which influence outcome expectancy as well as self-efficacy in relation to the hazard event (Paton, 2003). Place attachment, for example, can negatively impact disaster preparedness or even acknowledgement of the risk of natural hazards (Adie, 2020). Bonaiuto et al.'s (2016) review of literature on place attachment and natural hazard risk shows that both positive and negative relationships exist between place attachment and risk coping. Strongly attached individuals perceive natural environment risks but underestimate their impact and tend to be unwilling to relocate in the face of such risks but are more likely to return to risky areas after a disaster.

Though, place attachment has an effect on disaster risk awareness and mitigation behaviours (Miceli, Sotgiu, & Settanni, 2008), it must be recognised that there are other multi-layered influences on disaster preparedness ranging from internal factors (psychological), external factors (peers and neighbours), and contextual factors (community self-efficacy) (McIvor et al., 2009). Some communities and households are more capable of dealing with hazard events and the associated disparate outcomes due to the amount of physical and social resources they have access to for disaster preparedness (Norris et al., 2008). Preparedness can be thought of as consisting of four related activities: provisions (storing food, water, batteries, medical kit, and other supplies), skill level (ability to carry out first aid, emergency house repairs, and extinguishing a fire), planning (having an evacuation plan, building an underground shelter), and protective actions (access to a shelter) (DeYoung & Peters, 2016; Kirschenbaum, 2002). It is, therefore, important to consider the social and community context of each household and individual when assessing preparedness behaviour (DeYoung & Peters, 2016). This also requires consideration for groups such as international and domestic visitors that may be unfamiliar with not only the types of disasters that can occur in a place but also the preparedness behaviours expected from them.

It remains, however, unclear how sense of place affects recovery or willingness to return to affected areas due to the contradictory findings in existing studies (Chamlee-Wright & Storr, 2009; DeYoung & Peters, 2016; Silver & Grek-Martin, 2015). There has been several call for attention to the role of sense of place in disaster recovery, as well as the role of place identity and dependence in motivating people to return and attempt to rebuild their communities (Bonaiuto et al., 2016; Chamlee-Wright & Storr, 2009). Typically, post-disaster reconstruction initiatives tend to focus on infrastructure rather than people and a sense of place, which is actually crucial for long-term recovery of affected communities (Zetter & Boano, 2010). The challenge still remains to preserve and restore the environment and sense of place post-disaster (Chamlee-Wright & Storr, 2009). Issues of place making, the likelihood of returning to a destroyed place, place attachment–detachment are all relevant to the post-disaster context (McKinzie, 2019).

Existing studies show both positive and negative relationships between sense of place and disaster recovery initiatives. On the positive front, place attachment can be stronger among residents who have lived in a location for a long time and when community involvement is high (Shin, Nakakido, Horie, & Managi, 2016), and therefore, residents are more willing to adapt and adjust post-disaster to both physical and social changes to place. There is evidence to suggest that a strong sense of place can make communities more resilient post-disaster (Boon, 2014), and this can improve the attractiveness of the place for visitation by outsiders as they may want to see how a place is bouncing back and moving on. An awareness of what residents value in a place is heightened amid the recognition of loss and the changing nature of place as a result (Chamlee-Wright & Storr, 2009). Therefore, it is necessary for post-disaster place rebuilding initiatives to consider what aspects of place contribute to place identity and sense of belonging, given that place attachment can help people recover more quickly (Cox & Perry, 2011). Disasters can foster a strong sense of place and place attachment stemming from shared loss (Milligan, 1998), and therefore, intangible losses of identity, community, and sense of place may be of greater significance than tangible losses for those affected by disasters (Alston, Hargreaves, & Hazeleger, 2018). Cloke and Conradson (2018) showed the importance of community initiatives led by transitional organisations post-disaster to foster a sense of community and rebuild social connections following the Canterbury earthquakes.

On the negative front, residents can express distress over changes in physical and social landscapes (Cox & Perry, 2011), which impact their sense of place.

Disasters can lead to a decline in place attachment due to sites that are meaningful to individuals being destroyed or the replacement sites not aligning with individual and community expectations (Chamlee-Wright & Storr, 2009). Community attachment was correlated, for example, with negative affect following the BP Deepwater Horizon disaster (Lee & Blanchard, 2012). In the study of McKinzie (2019) the loss of the physical environment due to tornadoes was viewed by participants as a loss to the sense of self. Following disasters, often priority is given to rebuilding of infrastructure with limited consideration for psychosocial recovery, which is linked to sense of place. Psychosocial effects that shape people's capacity to cope and adapt, or maladapt is linked to feelings of belongingness and disorientation (Alston et al., 2018; Silver & Grek-Martin, 2015). Loss of surroundings after a disaster can result in residents feeling incredulous, unbalanced, frenzied, and full of negative emotions (McKinzie, 2019). They cannot make sense of themselves; the environment and their assumptions of a safe society are called into question (McKinzie, 2019). Disruptions can, therefore, bring to fore the unconscious meanings and significance of place to individuals (Chamlee-Wright & Storr, 2009). In this reflection, there is an opportunity for recovery initiatives to address the loss of sense of place by integrating changes to place in the wider discourse of more sustainable, resilient, and equitable societies. This can buffer the feelings of loss and recast perceived changes to place as contributing to the betterment of society.

Few studies have explored the nexus of place attachment and tourism in relation to disasters. A conceptual framework by Adie (2020) examined the relationship between place attachment and post-disaster decision-making in the context of second home owners. Based on interviews with tourism business owners and tourists on Hurricane Harvey that swept across Texas in August 2017, Zavar et al. (2020) highlighted the importance of social networks and place attachment in bringing tourists back to affected areas. Both place identity and place dependence were shown to have a positive effect on community resilience in earthquake affected tourism communities in Sichuan Province, China. These effects hold true for both residents and lifestyle immigrants (Guo et al., 2018). These studies confirm the role of sense of place as an important determinant of tourists in returning to disaster-stricken places and the role place attachment in facilitating disaster recovery and destination resilience through establishing emotional connections to places among potential visitors.

Box 7.1 Servicescape and place attachment after the Canterbury Earthquakes

The city of Christchurch in New Zealand was impacted by a magnitude 7.1 earthquake in September 2010, followed by a magnitude 6.3 earthquake in February 2011. With the second earthquake having an epicentre close to the city, this quake resulted in 185 fatalities (Hall et al., 2016). Thousands of residents were forced to leave their homes and the senses of place for many had to be renegotiated (Cloke & Conradson, 2018). Not only impacting sense of place, the Canterbury earthquakes also disrupted the imaginaries that had previously informed many individual and collective narratives of the city as a genteel, neo-gothic, English settler city, with significant cultural and landscape attractions (Cupples & Glynn, 2009). What was promoted in tourism terms as 'the garden city'. The damage and subsequent demolition of 1200 buildings in the city centre forced significant changes to existing patterns of urban mobility and interactions (Cloke & Conradson, 2018). The planned redevelopment of Christchurch city centre is a classic case of top-down place making (Amore & Hall, 2016, 2017). In parallel, several transitional organisations have emerged post-quake to curate sites and spaces for shared action, for promoting social reconnection and creating an aesthetically connected buzz to bring back residents to the city (Cloke & Conradson, 2018). This highlights the importance of social capital post-disaster for a community to regain their sense of place. This case study therefore examines how the servicescape can contribute to international visitors making sense of the city post-quake.

Introduced by Booms and Bitner (1981, p. 38; cf. Hall, 2008b), the idea of servicescape has been conceptualised as 'the environment in which the service is assembled and in which the seller and customer interact, combined with tangible commodities that facilitate performance or communication of the service'. Initially characterised as aspects of a firm's environment, servicescapes are conceptualised as organisationally controllable physical stimuli that enable firms to influence customer perceptions and satisfaction with the provided service. Bitner (1992) extended the definition of servicescape into three environmental dimensions, namely (1) ambient conditions (e.g., temperature, air quality, noise, music, odour), (2) spatial layout and functionality (e.g., layout, equipment, furnishing), and (3) signs, symbols, and artefacts (e.g., signage, personal artefacts, style of décor). The total configuration of these physical elements constitutes the servicescape (Finsterwalder & Hall, 2016; Tombs et al., 2017). Following a disaster, all three aspects of the environment can be impacted in a negative way. For example, the destruction of buildings can alter the spatial layout and functionality of an entire city. As others have suggested, the loss of manmade and natural markers as well as the psychological disruptions related to loss of home, place, and identity can lead to disorientation for individuals (Cox & Perry, 2011). This is where the servicescapes can facilitate reorientation and familiarise residents and tourists with place-based changes following a disaster. Servicescape can be designed and reimagined to influence cognitive, emotional, and physiological responses to increase the desire to approach and reduce behaviours focused on avoiding a particular service setting (Tombs & McColl-Kennedy, 2003; Tombs et al., 2017). However, very little attention has been paid to exploring the temporal dimension in connection with servicescapes (Finsterwalder & Hall,

2016) and attachment. In particular, how place attachment is maintained during the transition phase between the removal of, or disruption to, one permanent servicescape and the reestablishment of its replacement remains sparsely researched (Tombs et al., 2017).

Numerous new buildings and anchor projects have been or are under construction in the city centre (e.g., justice and emergency services precinct, performing arts precinct, and Avon river precinct) as part of the rebuild in the years following the earthquakes. For example, the Avon river precinct has been imagined as an urban waterfront where shared cultural values are celebrated. The Christchurch City Council has committed NZ$30 million towards the development of the performing arts precinct, which includes the construction of The Court Theatre, a public realm area, and landscaping. These projects attempt to link different servicescapes across the city in a way that gives new meaning about the city centre to both residents and visitors. In-depth interviews with 21 tourists visiting the city in 2018 reveal several themes consistent with the disaster literature on how residents or visitors perceive place changes (see Cox & Perry, 2011; Silver & Grek-Martin, 2015).

The findings reveal, first, the concept of 'reorientation' and 'disorientation' (Cox & Perry, 2011) among visitors. For those who had been to the city before the earthquake there is a conflicting sense of place. As one participant said, the landmarks in the city serve as a reminder of the past and seeing the past in relation to the new (e.g., new buildings and new precincts under construction) allowed this visitor to better understand the identity of the city going forward. It helped to clarify in his memory how the city was doing post-quake and the extent to which the past is being integrated with the new to maintain part of the past identity of the place. For this participant, the past and the new help with reorientation in relation to the 'new' sense of place. In contrast, other visitors who had been to the city prequake mentioned that the loss of buildings and landmarks has led to a sense of loss and grief as they find it difficult to understand where to go. This exemplifies Cox and Perry's (2011) disorientation. The sense of loss is more prominent among visitors who had been to the city before rather than first-time visitors.

For first-time visitors to the city, several of them mentioned that the servicescape is difficult to understand. As one tourist mentions,

> it [city centre] is not really centralised so all things are in a different place maybe because of the earthquake and the reconstruction but I feel like, usually in Europe we have cities where everything is in the city centre and here it is like everything is far away from each other.

This quote reflects the struggle that the visitor experiences to comprehend place identity in relation to other places she has visited and how the rebuild does not provide her with a clear sense of place. For other visitors, elements of surprise (negative) can be seen in the transcripts as they thought that the rebuild and disaster recovery activities would be at a more advanced stage seven years post-disaster. The landscape (damaged buildings and roadworks) contribute in many ways to reinforcing perceptions that the sense of place is still being imagined. In several of the transcripts, the servicescape can generate affect and consistent with Cloke and Conradson (2018) study, the affective landscape of transition has elements of positive (e.g., hope, renewal, and joy) and negative emotions (e.g., sadness, grief, and loss). Some perceive the new buildings and precincts as signifiers of progress, achievement, and determination of a community to create a new sense of place. For others, the new sense of place distracts from the original meaning of what the city was all about. Thus, the sense of belonging for even visitors can be laden with conflicts between what they imagine the place to be and what it actually is. The visitors seem to be renegotiating identity, which is often what residents experience post-disaster (Cox & Perry, 2011). These thoughts, perceptions, and attitudes in relation to sense of place can determine whether these visitors return or not.

Sense of place and continuous changes as a result of migration

As the various discussions throughout this book have highlighted, the relationship between migration and sense of place is complex. Migration is both an economic and emotional decision (Clark & Lisowski, 2019). Migration should be perceived as an ongoing process without an end point, as it is related to negotiations on life course events, and lifestyle aspirations, among others (Berg, 2020). It requires a fundamental commitment to changed workplaces, new social networks and relationships, and new residential environments (Clark & Lisowski, 2019). The tourism–migration nexus has received attention in tourism studies, with three clear research strands emerging. First, existing studies examine migrants' adaptation to their new environment by, for example, examining community integration of migrant tourism workers (Sun, Ling, & Huang, 2020) and their tourism-related entrepreneurial behaviours (Xiong, Zhang, & Lee, 2020). With a few exceptions, for example, Janta and Christou (2019) studying the hosting practices of migrant women, the hospitality practices of migrants have been so far neglected. There is also an implicit assumption in many studies that these migrants have arrived legally in their country of adoption. Recently, Aparna and Schapendonk (2018) challenged this notion by examining guest–host relationships and hospitality practices among asylum seekers. The second strand examines the tourism practices and behaviours of migrants within their country of adoption or on their travels to their home country. For example, studies examine the long-term visitor practices, e.g., second home owners (Åkerlund, Lipkina, & Hall, 2015), and diaspora tourism experiences (Li, McKercher, & Chan, 2020), tourism flows due to migration (Provenzano, 2020), and migrant identity, and travel behaviour (Yankholmes & McKercher, 2020). In this strand of research, tourist behaviour is examined under the lens of diaspora tourism, roots tourism, or VFR (visiting friends and relatives) tourism (Huang, King, & Suntikul, 2017). More importantly, migrant tourist behaviour is portrayed generally as a positive behaviour by focusing on the economic and social value of the tourism activities they undertake. The third strand examines the role of tourism sites in facilitating migrant adaptation to the new environment or as a way to reverse migration trends. For example, tourist sites of global importance have been shown to reverse city migration and bring positive social change (Buckley et al., 2020). Less clear in each of these strands is how sense of place informs these tourism activities, practices, and behaviours.

In the broader literature on sense of place and migration, on the one hand, migrants return home or to their ancestral lands because of a strong sense of place related to ideas of origin, identity, kinship, custom, health, and prosperity (Myat Thu, 2020). This informs some of the migrant tourist behaviour studies. On the other, the broader literature examines whether migrants feel a sense of belonging or attachment to their 'new' country and communities. This relates to broader ideas of assimilation and multiculturalism and can be linked to the migrant adaptation studies in tourism. For example, Wu and So (2020) investigated whether sense of belonging of immigrants (national belonging) to Canada improves when ethno-racial heritage belonging (ethnic belonging) is stronger. They found that increases in ethnic belonging significantly predicted increases in national belonging for both ethno-racial minorities and Caucasians. Beyond attachments to landscape and values, migrants return home also for reasons related to spirituality and well-being given that maintaining family ties can be associated with ancestral protection (Myat Thu, 2020). Motivations, experiences, and understandings of 'return' to origin places are highly personal, gendered, and generational with these return mobilities demonstrating often the agency, adaptability, and resilience of conflict-affected populations (Myat Thu, 2020). Clark and Lisowski (2019) also highlight the importance of examining the intersecting roles of place attachment, risk aversion, and social capital in decision-making models of mobility.

Still others look at migration from rural to urban areas within a country and the corresponding sense of place effects. This can have implications for the way domestic tourism is represented and marketed to such migrants. For example, Husa and Morse (2020) found that those who grew up in America's rural areas have a different set of values and social as well as place-based attachments compared to their urban peers. This attachment seems to emanate from the landscape, culture and hospitality, general contentment with living in a rural area, family connections, and the desire to raise children in such an environment. Therefore, value-based marketing appeals for tourism products can potentially work for this segment. A longitudinal study of urban–rural migrants within Norway showed that well-being and place attachment are key factors influencing the decision to stay in or leave the countryside. The study pinpoints to the importance of social contacts, materialities, the past and memories, and emotions and affect as being significant in the consideration for staying or leaving the countryside (Berg, 2020). Examining migrants from rural areas sense of belonging to host cities in China, Huang et al. (2020) found that contact with residents who were not related to them was positively linked to the development of a sense of belonging, while contact only with non-residents in these host cities had the opposite effect. Length of residence is a significant factor in determining place attachment for migrants. A study by Counted, Moustafa and Renzaho (2019) shows that sub-Saharan African residents living in Netherlands had stronger place attachment and place dependence if they stayed more than five years compared to those who stayed for less than five years. But beyond length of residence, education levels, gender, place of residence, and language proficiency in Dutch had some effect on place attachment, or place identity and place dependence. This highlights the role of socio-demographics in determining type and strength of attachment for migrants.

Du (2017) explores how geographic mobility is implicated in the process of place bonding. By examining different types of young adulthood migrants and returnees, the study challenges the conservative view of portraying place attachment and belonging to homeland as universal and unconditioned. The results suggest that geographic mobility does not undermine place attachment or sense of belonging but tends to attenuate its intensity. Place attachment is not necessarily bounded by a single place as homeland but place attachment ascribed by birth can be stronger than attachment acquired by residence. This has been noted in other studies, where even connotations of an ancestral home did not have the same meaning for migrants from the same country. For example, Li and Chan (2018) investigated Chinese migrants' place attachment to China from a multi-generation perspective. The results show that meanings of ancestral home have place, person, and process dimensions and three groups with different types of connections were identified. Those that were personally attached to China as the ancestral home strongly valued personal involvement through the physical and social ties they had. Those who were cognitively connected to a generic China and Chinese culture were more aware of the influence of their own families and the Chinese communities and embedded their connections in understandings and knowledge of Chinese cultural practices. The third group was spiritually connected to a symbolic China, where this connection is maintained through collective memories, sense of pride, and appreciation for family ancestry and interests in their personal past. The findings suggest that the meanings of an ancestral home are diverse and dynamic, reflecting both individual and family's migration history, personal experiences, cultural exposures, social relationships, and mobility (Li & Chan, 2018). Cohen, Duncan, and Thulemark (2015) use the term 'lifestyle mobilities' to challenge discrete notions of 'work' and 'leisure' and that of 'home' and 'away', in a quest to show the increasing fluidity between travel, leisure, and migration. The study shows the complex relationships between issues of identity construction, belonging, and place attachment associated with sustained corporeal mobility (Cohen et al., 2015). Other terms that occur in the literature that highlight the interelationships between tourism-related mobilities and place attachments include that of multiple dwelling and dual dwelling (Hall,

2014b; Hiltunen et al., 2013). Next, the example portrayed in Box 7.2 illustrates Chinese migrants' sense of place and tourist behaviour.

Box 7.2 Chinese migrants in the US – sense of place and tourist behaviour

There are different facets of 'China' that migrants are attached to (Li & Chan, 2018). Nonetheless, many of them undertake the journey home as part of what can be viewed as diaspora tourism for several reasons including leisure, local food, culture and heritage, social connections, personal identity, and family obligation (Li et al., 2020). Chinese migrants in particular tend to travel back home during the Golden Week (Chinese New Year). For Chinese migrants in the US, their connection to homeland can be seen from various activities they undertake. For example, Huang and Chen (2020) found that these migrants in their adopted country connect with 'China' through participation in, for example, Chinese cultural activities and join Chinese ethnic clubs/organisations. These reflect practices that are centred on cultural events. These migrants also keep in touch with homeland through watching Chinese movies, TV shows and listening to Chinese songs and music, shop at Asian groceries, and eat Chinese food. In essence, the activities help to maintain a sense of identity and community for those migrants. Similar to the study of Li and Chan (2018), inter-generational differences were investigated and the results showed first-generation migrants and those who had been in the US for more than four generations had higher scores on these activities (Huang & Chen, 2020). Of particular significance is that first-generation migrants travel more often to China than other generations surveyed but then overtime, the propensity to travel to China increases for those who are fourth generation and above. These results suggest that there are ways of belonging that are rooted across generations and Chinese migrants get a sense of place of their homeland through the performance of various activities in their adopted country and travels to China. Both place identity and place dependence are at work among these migrants in relation to their homeland.

Sense of place and incremental changes as a result of environmental change

Not only do people construct the environment but the environment constructs people, which reflects the social construction of places in relation to the natural environment (Barrios, 2011; McKinzie, 2019). What is considered in nature or the environment varies by time, place, and culture (McKinzie, 2019). Research on sustainability and environmental change recognises the role that sense of place and place attachment plays in shaping framings of environmental issues and motivating behaviour change (Devine-Wright, 2013). In particular, place identity and place attachment are increasingly recognised as contributing to the understanding of human responses to environmental change. Differences in sense of place can lead to individuals to react differently to potential environmental threats to their communities (Kaltenborn, 1998). Place attachment can engender pro-environmental behaviours (Zhang, Zhang, Zhang, & Cheng, 2014). It can also increase advocacy for mitigating climate change impacts on national parks and increase willingness to engage in climate friendly behaviour at such parks (Groshong et al., 2020). Evidence of the importance of place attachment as a means for engaging visitors in climate-related actions both with and beyond a destination exists (Groshong et al., 2020).

Albrecht (2005) uses the term 'solastalgia' to describe feelings of distress and anger experienced by people who had lost their solace and place attachment towards their home and territory because of environmental degradation, including climate change and its effects on human security in a tourism context (Hall, 2015). For example, Marshall et al. (2019) attempt to understand whether degradation to places and threats to ecosystems affect the meaning and well-being of different groups (e.g., residents, tourists, fishers, and tourism operators) in relation to coral bleaching and mortality at the Great Barrier Reef in Australia. They argue that individuals experience ecological grief, similar to solastalgia, if they ascribe meanings or intrinsic values (such as aesthetic, scientific, or biodiversity-based values) to the place and that such meanings are centred on concepts such as place attachment, place identity, lifestyle dependence, place-based pride, and the derived well-being. The findings of the study show that 'Reef Grief' is closely and positively associated with place meanings for resident and tourist populations, having significant implications for how individuals and populations experience ecosystem decline and loss within places that are meaningful to them (Marshall et al., 2019).

The literature converges around the importance of sense of place and place attachment in tackling issues related to climate change (Singh et al., 2020). For example, studies on place attachment and climate change adaptation have taken two approaches of either examining how place attachment can be a motivator for adaptation behaviours or how place attachment becomes a barrier to transformative adaption (Clarke, Murphy, & Lorenzoni, 2018). Devine-Wright et al. (2015) show the importance of different scales of attachment (global vs. local) in determining whether individuals are more likely to attribute climate change to anthropogenic causes or not. Unless a place is significantly disrupted by climate impacts, residents are often not willing to adopt behaviours that can mitigate these impacts as shown in the studies of Hall (2006b) in terms of flood events in New Zealand and Clarke et al. (2016) in relation to flood risk in Ireland. Clarke et al. (2016) found that transformative strategies failed to materialise because of socio-cultural values related to place attachment and identity, institutional reliance on technical expertise, which fails to look beyond traditional technocratic approaches, and institutional regulatory approaches. By analysing attitudes and adaptation to climate change in communities of the geographically remote and climate-vulnerable Outer-Hebrides in Scotland, Butts and Adams (2020) showed that changeable, difficult and unpredictable weather binds people to place and influences how they think about themselves, their place, and adaptation in dynamic ways.

It is, therefore, not surprising that perceptions of the impact of climate change can contribute to mass migration but the actual impacts can also contribute to disasters. Place attachment has been shown to be a factor in increasing the adaptation of individuals who are impacted by climate change to changes in their surrounding environment (Khanian, Serpoush, & Gheitarani, 2019). Looking at such effects in Western Iran, Khanian et al. (2019) found that perceptions of impacts on the environment were related to place attachment, which also contributed to discourage migration. Therefore, a balance exists between subjective adaptive capacity and climate change impact perception in which the place attachment factor can tilt the balance in favour of more adaptation and less migration. Place identity can increase climate friendly behaviour in relation to advocacy actions that imply long-term engagement with national parks while place dependence increases both climate friendly behaviour related to the short-term, immediate actions taken at the park during the visit as well as support for climate friendly management action (Groshong et al., 2020). Sense of place can motivate stewardship actions towards the local natural environment and promote sustainable behaviours (Chapin III & Knapp, 2015). However, these effects are not uniform across contexts and population groups. For example, tourists are uninformed and remain apathetic to climate change impacts on destinations as noted in several studies (Dillimono & Dickinson, 2015; Scott et al., 2012).

Globally, climate is changing and influencing where and when visitors decide to travel. Visitors' attachment can affect their intended future visitation under changing climate conditions. Investigating visitation behaviours to Mount Desert Island (MDI), Wilkins and de Urioste-Stone (2018) found that visitors with a high level of place attachment participated in more recreational activities during their trip. Additionally, those with high attachment said they were less likely to be deterred from visiting MDI in the future under potentially negative changing environmental conditions (Wilkins & de Urioste-Stone, 2018). Kajan (2014) investigated place attachment in tourism-dependent communities in the Arctic through place dependence and place identity. Tourism in the Arctic is increasing rapidly and affecting local communities through land-development pressures. As the impact of climate change worsens for such areas, this study revealed that attachment was expressed in two different ways: economic activities manifested almost exclusively through tourism development and non-economic activities present through consumption-related actions (e.g., fishing) as well as through less tangible issues such as connecting with nature. The economic and consumption-related activities were attributed to place dependence, whereas the more intangible factors with place identity (Kajan, 2014). Box 7.3 illustrates the impact of environmental change on a native species, glow-worms, in New Zealand and the implications for sense of place.

Box 7.3 Glow-worms at Waitomo Caves, environmental change and sense of place

Native species can not only contribute to destination attractiveness but also allow visitors to develop a sense of place. Usually associated with the Waitomo Caves in New Zealand, snare-forming glow-worms (Arachnocampa spp.) are an important element of tourism in New Zealand (Hall, 2013). Similar to the kangaroo, koala and wombat in Australia, the Tasmanian devil in Tasmania, and the kiwi bird in New Zealand, these native species often are not only associated with the identity of the destination but also allow visitors to develop a bond with the place. Native species have a well-defined natural range to which, by evolution, climate, and geography they are place bound (Forristal, Lehto, & Lee, 2014). Therefore, the loss of any native species can have a dramatic impact on the ecology of a place and subsequent impacts on the tourism industry. As an example, the loss of coral species and fish at the Great Barrier Reef has already prompted the Australian Government and tourism industry operators to rethink how the attraction is marketed.

Hall (2013) notes that glow-worms can be affected by tourism in many different ways through direct disturbance, impacts on the number and type of glow-worm prey, changes to glow-worm habitat, and environments. Impacts can also be understood from the perspective of the natural habitat itself. Glow-worm viewing is an activity that occurs in a cave or in a rainforest at night-time. A report by Kenny et al. (2001) showed that the Waitomo GlowWorm Cave would be closed more frequently owing to flooding, and siltation in the cave would increase as climate change impact becomes more prominent. As visitor numbers rise, keeping the caves intact and the glow-worms alive is critical to the industry's survival. The first wakeup call was in the 1970s, when the glow-worms just turned off their lights and the cave was closed for about nine months. Scientists have since learned a lot more about glow-worms and the climatology and ecology of their subterranean homes. Since then, the cave environment is closely monitored with data on water levels within the cave, and alarms go off if a flood is likely. Carbon-dioxide levels are also monitored given that they are influenced by visitor numbers and climatic conditions, and high levels can also damage the cave's iconic limestone formations (Evans, 2017). Therefore, monitoring environmental change goes hand-in-hand with conservation efforts for both the native species and their natural habitat. Glow-worms and the Waitomo caves are tourism resources of national significance, for which the Government and communities, on behalf of the people, have a major custodial responsibility. Increasing tourism presents a growing potential for conflict between the dual requirements of presenting the resource and at the same time protecting it (De Freitas & Schmekal, 2006), with implications for not only locals' sense of place but also how tourists make sense of their visit to the caves. The glow-worms are Waitomo's and even New Zealand's point of difference, therefore, conservation efforts should be prioritised at the expense of economic return as there is no guarantee that without such an approach, the lights will still be on literally!

Conclusion

This chapter highlights sense of place as a dynamic concept that changes and evolves as a result of place-based changes. In particular, the influence of three types of changes is examined (sudden, continuous, and incremental). These changes can not only change the physical characteristics of place but also contribute to redefine the socially constructed meaning of place for different groups. As such, notions of place identity and place dependence for different groups (locals and tourists) cannot been seen as static but rather as fluid concepts that evolve as a result of disasters, migration patterns, and environmental change. Sense of place is not only affect laden but this affect can be negative due to perceived impacts of change on place. The chapter shows that sense of place has a significant role to play in disaster preparedness and recovery but can also induce migration. Likewise, migration can be seen as searching for sense of place in an adopted country but also as seeking rootedness through travels to homeland. Though environmental change can precipitate both disasters and migration, it is clear that sense of place can act both as a motivator for and barrier to adaptation for individuals and communities to such changes.

8 Exploring the virtual world

The sense of place from VR tourism experiences

Introduction

Virtual environments (VEs) and virtual worlds (VWs) are often associated with the ability of the medium to convey the desired information and experience to the user (Lombard & Ditton, 1997). The medium is usually immersive technology that blurs the boundary between the physical and VW (Suh & Prophet, 2018). VWs can be defined as 'online environments for people to access information and have experiences with the added feature of an awareness of others' (Li & Maher, 2000, p. 103). VWs can create sense of place through metaphorical reference to physical places, and often complex and personal spaces (Li & Maher, 2000). However, reference to physical place is not necessary for a user of VWs to feel presence (Plunkett, 2011). For example, online VW technologies such as Second Life and OpenSim, and gaming technologies such as World of Warcraft and Unreal Tournament (Falconer, 2017), enable movement and interactions in VWs through the agency of an avatar, with any similarity to physical places not required to feel immersed.

> The arrival of virtual reality (VR) devices, 4D movies, simulation rides, and video conference environments provide users with a rich media experience. Combined with naturally realistic environments and immediate feedback, the mediated environment feels as if it is not mediated (Wang & Hsieh, 2020, p. 696).

Terms such as VEs, VWs, and VR are not exactly the same (Yung & Khoo-Lattimore, 2019), with VR being the most immersive form of VWs (Singh & Lee, 2009). A distinction is also made between VR and artificial reality (AR), with the latter allowing for 'the superimposition of synthetic images over real images providing augmented knowledge about the environment in the user's vicinity which makes the task more pleasant and effective for the user, since the required information is spatially superimposed over real information related to it' (Danado et al., 2003, p. 1). Thus, AR allows users to interact with the real environment while VR relies more on the VE (Liang & Elliot, 2020). Mixed reality (MR) bridges the divide between AR and VR and can be considered as a space in which both real and virtual objects are presented simultaneously within a single display (Milgram & Kishino, 1994).

Already a substantial knowledge base exists that views online environments as 'places' (Moore et al., 2009; Paiva, 2015) and argues that presence allows for sense of place to emerge (Paiva, 2015; Plunkett, 2011; Steinkuehler & Williams, 2006; Turner & Turner, 2006). In virtual spaces, the sense of being there has been referred to as immersion, embodiment, telepresence, or simply presence (Bryson, 1995; Gutierrez et al., 2008; Paiva, 2015). Steuer (1992, p. 76) defined presence (telepresence) as 'the extent to which one feels present in the mediated environment, rather than in the immediate physical environment'. Feelings of telepresence are naturally subjective and are influenced by a VE's ability to engage the users' senses during the experience (Gutierrez et al., 2008) and can be described as a mental experience affected by different conditionings (Paiva, 2015). VEs can be described by their capacity to provide physical immersion and psychological presence (Gutierrez et al., 2008). This immersion can influence a user's feeling of presence or telepresence (Baños et al., 2004) and refers to the extent to which a user is isolated from the real world (Gutierrez et al., 2008; Guttentag, 2010). Others suggest that presence is dependent on factors such as interactivity and spatial presence (Bryson, 1995). Evoking a sense of place in a simulated environment should, thus, be triggered by feelings of presence, interactivity, flow, and immersion to provide supporting interactions for users (Animesh et al., 2011; Bryson, 1995; Ghani et al., 2019). However, a high degree of realism is not a necessary condition for sense of place in mediated experiences (Turner et al., 2013).

This chapter examines virtual spaces as places and the corresponding sense of place, with a focus on VR rather than AR or MR. The chapter starts with a discussion of the characteristics of virtual places that trigger sense of place or place attachment. This discussion then progresses to examine the state of the art on VR and tourism. Finally, the chapter outlines the factors that influence tourists' sense of place in VR, given that telepresence has a direct influence on consumer attitudes and behavioural intentions (Dholakia & Zhao, 2009; Fiore, Kim, & Lee, 2005), thus affecting repurchase decisions for virtual tourism experiences. VR can also have higher effects on tourist emotional responses (Yeh et al., 2017), thereby contributing to higher levels of attachment compared to physical places. Tourists' can be attached to VWs and the experience provided in them (Plunkett, 2011; Turner et al., 2013), but attachment to the physical world can be due to the VW as well (Wang & Hsieh, 2020). Therefore, the relationship between VR, sense of place, and tourism is complex and dynamic.

Virtual space, place, and sense of place

How people conceive of place varies immensely, and to a certain point all places are imaginary for they exist in the mind and in the field. The boundary between physical space and virtual space is blurring. Virtual spaces have become, culturally, actual spaces given that they have been assimilated as part of everyday life (Paiva, 2015). Interacting in an online community, using your own avatar, while sitting in a café illustrates this blurring boundary. Relph (1976) suggested that place can be thought of as space and meaning and therefore, place identity has three components, namely, physical setting, activities afforded by the place, and meanings attributed to the place. A vibrant online community can provide exactly that, which outlines that placeness can exist in VR by giving spaces multi-layered social meaning (Nitsche, 2008; Paiva, 2015). Space has a key role in framing an understanding of narrative meanings (Kukkakorpi & Pantti, 2020), and VR provides this opportunity through interactions with the environment itself and the constituent objects. The sensation of being there has, therefore, been a key concept for understanding and evaluating the effectiveness of VR. For example, visualisation and design aspects, including soundscape have been shown to impact on user's sense of place in VEs (Falconer, 2017). Sensory perceptions, awareness, and meaningful interactions can enable place attachment in VWs (Goel et al., 2011). A state of deep involvement (called cognitive absorption) that users experience as they perform an activity that results in them losing track of time can enable place attachment. Awareness of who they interact with and how they interact, what they interact about, and where in the virtual sense this interaction occurs can also contribute to sense of place (Goel et al., 2011). Virtual spaces can become symbolic containers of memory as users spend time in them, providing a sense of place much like real places (Nitsche, 2008). Yet, it must be recognised that real places are often richer and more emotion laden than spaces created by technology and also exposure to real places tends to be significantly longer before place attachment develops while a typical VR experience is usually shorter in duration (Turner & Turner, 2006). This implies that sense of place as conceptualised by sociologists and human geographers may not have the same foundation and intensity of connections in VR or VWs.

Evidence exists to suggest that users of VR and VWs can develop a sense of place (Goel et al., 2011; Nitsche, 2008; Paiva, 2015; Turner & Turner, 2006), despite the particularities of the experience in such spaces (Paiva, 2015). Relph (2007b) defines virtual geographical place as those that capture the manifold qualities of the real world. Despite Relph (2007b) considering virtual space as a reproduction of actual places, there can also be no real world components that imbue a virtual place with qualities (Paiva, 2015). As an example, community ties based on diaspora may exist only in virtual space and not materialise in real-life (Paiva, 2015). VR can establish its own conventions and people adapt themselves to them (Paiva, 2015), but it is through meaning ascribed to such practices that sense of place emerges. VWs ignite interest of the mass as technologies that can provide enhanced opportunities to produce virtual spaces where meaning grows through shared experiences and interactions with the online environment to give a resulting sense of place (Falconer, 2017). Participants' direct observation of VWs and their interpretation of the experience in such spaces in real time allow them to make sense of the experience (Ghani et al., 2019). This process requires consideration of the physical attributes of the virtual space, including non-visual senses (e.g., soundscapes). It also requires consideration for activities that users will engage in, whether actual or desired (Turner & Turner, 2006), the social interactions they will have in the space (e.g., through avatars), and the immersive device factors. The latter, refers to the effects of the device involved, towards the participant's' experience of virtual place (Turner & Turner, 2006). Place attachment can emerge from the meaningfulness of the interaction itself rather than the physical or virtual space in which it occurs (Goel et al., 2011).

Tuan (1977) emphasised the role of objects and their symbolic meaning in the constitution of the 'feel' of place. The 'object' in VR can be multi-fold given that the features of the virtual space itself can have an effect on the experience (Paiva, 2015). Factors that influence presence in VR include extent and fidelity of sensory information, including vividness; match between user actions and display; content factors including objects, actors, events, interactivity, autonomy in action, reaction of others, nature of task, meaningfulness; and user characteristics (IJsselsteijn et al., 2000). Goel et al. (2011), for example, describe the salient attributes of VWs that compel users to return. These attributes comprise of aura, focus, nimbus, and boundaries, all of which can be altered to influence the user. Aura is the mechanism by which the medium (audio, visual, or textual) in the VE comes to being in relation to an object (Benford et al., 1994). Focus is the ability to delimit an observed object's interest such that the more an object is within a user's focus, the more he or she becomes aware of it (Benford et al., 1994). Nimbus is the ability to represent an observed object's projection towards the user or the avatar such that the more an object is said to be within your nimbus, the more the user is aware of it (Benford et al., 1994). Aura, nimbus, and focus are properties that allow for the description of interactions between users, and between users and objects, in terms of their awareness of each other (Goel et al., 2011). Boundaries, to the contrary, divide space such that they provide mechanisms to mark territories, control movement, and influence interactional properties (Benford et al., 1994). Thus, users intend to return to VWs after having conceived of it as a 'place' in which they had meaningful experiences (Goel et al., 2011). The tracking of user's movements and direction of gaze can convey a sense of place (Turner et al., 2013).

Presence in virtual spaces does not, however, automatically create sense of place (Paiva, 2015). Mediated environments need to tell compelling stories and provide more than just a setting to interact and respond to the surroundings (Kukkakorpi & Pantti, 2020). For example, the aesthetic and symbolic qualities of the environment focusing on its potential to evoke emotional responses are critical for sense of place (Kukkakorpi & Pantti, 2020). In this regard, the distinction between spatial and emotional immersion is significant (Zhang et al., 2017), with the former referring to the engagement with the online physical environment, and the latter the engagement with the characters. Yet, this categorisation is not without tension because both the spatial qualities of the VE and the experience of it can create an emotional bond (Kukkakorpi & Pantti, 2020). In examining Second Life residents, Animesh et al. (2011) found that interactivity had a positive impact on both telepresence and flow; sociability, which fosters interactions with other participants, had a positive impact on social presence. Stability helps users to develop strong social bonds, which positively impacts social presence. Both social presence and telepresence affect flow, which is the perceived sense of enjoyment obtained from interacting in the VE, positively (Animesh et al., 2011). Paiva (2015) also notes that addictiveness to VEs can also inform user's sense of place (Paiva, 2015). This is particularly relevant to the online gaming world; dependence on the gaming environment can inform self-identity and give meaning to actions and behaviours in the real world. Thus, despite these studies not examining what determines sense of place in VEs, it seems that beyond telepresence, social presence, enjoyment, aesthetic and symbolic qualities, and addictiveness can contribute to a sense of place. In particular, addictiveness may be related to place dependence, while social presence would contribute to the social bonding aspect of place attachment.

While the previous discussion centres on the user immersing in the online environment to derive meaning, the online environment can also be designed to engage players in physical activities at real-life locations (e.g. Pokémon Go). Specifically, AR games such as Pokémon Go show that such experiences can transcend the virtual, spatial, social, and physical worlds (Clark & Clark, 2016). AR is a type of VR (Guttentag, 2010). By connecting the real world with augmented reality games, for example, can heighten the interactive and immersive experience, enabling more meaningful experiences (Wang & Hsieh, 2020). Location-based AR game characteristics that facilitate player's enjoyment can contribute to place attachment to physical locations (Wang & Hsieh, 2020). This is because co-presence in the real location augments the meaning derived from the online environment where the majority of the experience happens. Co-presence can be defined as the feeling of being together with another person in either a virtual or physical space (Slater et al., 2000). Co-presence has largely been used to understand social interactions with remote users in VR (Wang & Hsieh, 2020). By shifting co-presence to both the simulated virtual word and real world, the sense of spatial presence is heightened. This blend of the physical and the fictional world simultaneously (Wang & Hsieh, 2020) epitomises the blurring of boundaries between reality and VR. Box 8.1 illustrates how sense of place in VR can increase pro-environmental behaviours.

Sense of place and attachment to virtual worlds by tourists

The sense of 'being there' provided by VR (Steuer, 1992) has expanded tourists' information search behaviours, travel experiences, and post-travel behaviours (Guttentag, 2010; Yung et al., 2020). Tourists can now explore tourism destinations from inside their home and walk through a hotel room before deciding whether to book it. Travel agencies can use VR to create the 'illusion of non-mediation' (Lombard & Ditton, 1997) and immerse their customers into places they dream of visiting. Tourists may interpret information delivered through AR and VR as the information 'obtained from their own direct experience' (Klein, 2003, p. 42). AR applications are becoming more valuable for the tourism industry as they can increase social awareness of the immediate surroundings and allow tourists to gain a deeper knowledge of place (Martínez-Graña et al., 2013). In AR while observing real place, tourists can observe virtual objects overlaid on the real place usually through interactivity based on their own smartphones, tablets, and digital camera, see through displays, and spatial augmented reality (Loureiro et al., 2020), thereby enhancing the quality and authenticity of the experience. VR is

Box 8.1 Virtual reality, sense of place, and pro-environmental behaviours

Digital simulations of the natural environment can be seen as a method of heritage preservation (Le et al., 2005). This is particularly relevant to virtual heritage in the form of landscapes and scenery. Through VR experiences of such landscapes and scenery, individuals may not only form sense of place but also adopt pro-environmental behaviours to protect the actual landscape. This is shown below through an experiment that was carried out among visitors to a heritage site in UK to find out if, in the first instance, there is a higher presence from VR compared to exposure through pictures of the landscape. Second, the experiment was also designed to evaluate emotional responses and how much visitors were willing to give to a charity after the VR experience to preserve the actual heritage.

Based on 80 participants (between subject design), one group was exposed to pictures of the landscape (photo group) while the other group was immersed in a virtual experience of the landscape (VR group) using HTC Vive. T-test results showed that presence was higher for the VR group compared to the photo group on most of the telepresence items. The emotional responses from the VR experience were also higher compared to the photo group. For example, excitement (M_{photo}3.90 vs. M_{VR}5.20), curiosity (M_{photo}5.18 vs. M_{VR}6.10), and fun (M_{photo}4.18.90 vs. M_{VR}5.78) were higher for the VR group. The overall attitude towards the landscape was not statistically different between the two groups (M_{photo}5.10 vs. M_{VR}5.38) and intentions to visit the actual landscape were similar ((M_{photo}4.95 vs. M_{VR}4.63). Visitors had a higher sense of place in the VR group compared to the photo group, which led the former group having a higher willingness to donate money to a charity compared to other group. The donation was in the form of a bronze, silver, gold, and platinum donation to preserve the landscape and scenery. The VR group chose mostly the silver and gold donation categories while the photo group chose mainly the bronze and silver categories. These behaviours provide credence to the argument that VR experiences are immersive and have the ability to generate stronger emotional responses than traditional ways of conveying stimuli. The higher emotional responses and stronger sense of place seem to contribute to stronger pro-environmental behaviours, through for example, willingness to donate money.

also contributing to memorable tourism experiences (González-Rodríguez et al., 2020). Through immersive virtual experiences of destinations such as Seville and Barcelona in Spain, using PastView, González-Rodríguez et al. (2020) found that VR experiences can result in e-WOM recommendations on Trip Advisor, highlighting post-travel behaviours due to the immersive experience. It is, therefore, not surprising that the AR and VR literatures in tourism have increased exponentially with at least three recent systematic reviews of this literature (see Beck et al., 2019; Liang & Elliot, 2020; Yung & Khoo-Lattimore, 2019). Two notable issues identified are the lack of consistency in the terminology used (e.g., VR, VE, and VW) and theory to support findings (Yung & Khoo-Lattimore, 2019). Also, the literature on sense of place and place attachment emerging from tourists' VR experiences remains thin (Damjanov & Crouch, 2018; Pantelidis, 2019). The exception is the study of Kim et al. (2020b), which examines the factors that compel consumers to visit destinations presented in VR, with such visit intentions being predicted by attachment. However, attachment to VR as a technology rather than the visited destination was measured, suggesting that different forms of attachment should be considered when studying tourist populations. Tourists can attach to the technology itself (e.g., HTC Vive), the presented environment (destinations and tourist activities), stories/characters (e.g., avatars and other objects), and attachment to other participants/players. Moreover, related research by Kim and Hall (2019) showed the importance of the moderating role of being a visitor or non-visitor at the destination portrayed in VR tourism. Their results identified the highly significant effect of consumers' perceived enjoyment on flow state and of flow state on subjective well-being. Ongoing use of VR tourism was greatly influenced by flow state and subjective well-being. Importantly for destinations, the association between usefulness and flow state had a significant moderating effect depending on whether a visitor or non-visitor.

In tourism, VR and AR can provide a significant advantage over traditional tools of destination marketing such as travel brochures and other multimedia materials that give only a glimpse of a destination's attractions (Cheong, 1995; Liang & Elliot, 2020). Based on the level of immersion and technical capabilities, three experiences can be provided, fully-, semi-, and non-immersive VR (Beck et al., 2019). In semi- and non-immersive VR, the user still has some contact with the real world (Gutierrez et al., 2008). Fully immersive experiences are usually designed for a single user experience (Guttentag, 2010). Studies using fully immersive experiences are recent in tourism (e.g., Tussyadiah et al., 2018), with wired (e.g., Oculus Rift and HTC Vive) or wireless (e.g., Samsung Gear VR) Head Mounted Displays (HMDs) (Beck et al., 2019). Guttentag (2010) identifies several areas in tourism where VR can provide added value, from planning and managing the experience to its marketing. For example, the experiential nature of VR makes it an optimal tool for providing rich data to potential tourists in their information search on destinations (Guttentag, 2010). Yet, whether this search process through using VR can actually generate place attachment remains to be ascertained. VR's marketing potential can be linked to its ability to provide extensive sensory information (Guttentag, 2010). Stimulation of the senses can make a virtual place more real and has been suggested as a way to improve authenticity of virtual tourism experiences (Yung & Khoo-Lattimore, 2019). VWs are becoming increasingly realistic to the senses and therefore gaining meaning (Plunkett, 2011). This implies that the tourist can develop emotional bonds to such spaces. VR can also be used for tourist education, entertainment, and gamification, while also improving accessibility of destinations to disabled individuals (Guttentag,

2010; Yung & Khoo-Lattimore, 2019), but whether these foster a sense of place remains elusive. A 20-year review of research on VR and AR in tourism shows themes such as atmospheric design recommendations, cultural heritage and smart cities, location-based information and image quality, destination marketing, and telepresence and virtual communities are some of the areas where researchers have attempted to understand the influence of these immersive technologies on tourism (Loureiro et al., 2020). A review of the AR literature in tourism specifically shows emerging themes of AR design and development; user acceptance of AR; user experience, satisfaction and behavioural intentions; AR implementation and management; and gamification and AR (Liang & Elliot, 2020). Surprisingly, sense of place and place attachment do not appear as significant issues in both the AR and VR tourism literatures.

VR can cater specifically to the vicarious experience by allowing users to select the desired visual, audio, and spatial aspects of destinations without being there. This points to the mediated experience being related to presence as a significant factor (Tussyadiah et al., 2018). The concept of presence has been examined from different angles in the tourism literature (Tussyadiah et al., 2018). Being able to feel an actual atmosphere of the place through the sense of presence potentially can be far more engaging and convincing for tourists. Yet, presence and sense of place should not be confused, with the former relating to aspects of immersion, interactions, and flow that allow the user to feel telepresence, while the latter is an outcome that is dependent not only on presence but also other cognitive and affective stimuli. There are different types of presence. Presence has been conceptualised in terms of its descriptive (focuses on delineating dimensions) and structural (how presence is generated in the user's mind) models (Diemer et al., 2015). Others suggest three types of presence, namely physical, social, and self (Lee, 2004). The experience of presence is a complex, multi-dimensional perception, which is formed through an interplay of multi-sensory information and various cognitive processes (Diemer et al., 2015). Yet, there is no consensus on the dimensions relevant to tourism experiences. For example Hyun and O'Keefe (2012) use one dimension of feeling there, while others confirm that spatial presence, possible actions and self-location, social presence (Beck et al., 2019; Tussyadiah et al., 2018), and suspension of disbelief (Surovaya et al., 2020) are important dimensions to feel telepresence. However, telepresence has been shown to impact destination image, attitudes, and intentions to visit (Hyun & O'Keefe, 2012; Tussyadiah et al., 2018; Yung et al., 2020). For example telepresence has been shown to influence virtual cognitive, affective and conative image of destinations (Hyun & O'Keefe, 2012). A consideration for content quality, system quality, and vividness of the experience itself is also required to positively influence customers' attitude and telepresence, leading to their positive behavioural intention to visit the destination (Lee et al., 2020). Box 8.2 provides an example of how VR can be used in the family tourism context to reduce travel risk, improve destination image sense of place and increase intention to visit.

While feelings of telepresence can possibly foster sense of place, this cannot materialise without, for example, learning, familiarity, and emotions during the experience. These factors are significant to also place attachment in the real

Box 8.2 Family tourism, destination image, travel risk, and VR

Several studies have shown that VR can be used to minimise the travel risk and facilitate potential travellers' decision-making processes (Guttentag, 2010; Yung & Khoo-Lattimore, 2019). However, studies generally focus on individual VR experiences rather than group ones. In this study, 12 families experienced a destination through VR, using the Samsung Gear VR headset, and were interviewed after the experience (see Yung et al., 2020). The study sought to find out whether VR would have a positive influence on destination image evaluation and reduce perceived travel risk, and in this way increase their sense of place. Hyun and O'Keefe (2012) showed that VR has the ability to positively impact all three components of destination image (cognitive, affective, and conative). The results show that VR experiences affect cognitive image as family members have a greater capacity to recall attributes of the destination following the VR experience. This is related to the sense of being there, i.e., feeling of telepresence. This gives them a better sense of place before they travel. Though emotional responses such as 'amazing', 'wow', and 'love' were used to describe the experience, the affective component of image seems less prevalent compared to cognitive images (Yung et al., 2020). The level of immersion could possibly account for cognitive images being stronger and generating that feeling of 'place'. Certainly, the VR experience activated the conative component of destination image as participants expressed strong intention to visit the destination in the near future. Thus, elements of immersion, presence, and enjoyment were clearly at play in evaluations of destination image components. The VR experience specifically reduced perceived travel risk in relation to safety, performance, and financial risks. This example highlights VR as a destination marketing tool can be used to reduce destination choice risk, convey a sense of place, and improve pre-travel destination image as well as improve visit intentions.

world (Hosany et al., 2017; Prayag & Lee, 2019). Plunkett (2011) identifies several ways in which a visitor can become attached to a VW. Learning that occurs through virtual tours of hotels, destinations, and tourist activities creates a sense of familiarity with the technology, and therefore attachment can develop to the technology itself as well as the environments depicted in them. Through the use of maps, photos, videos, and verbal descriptions, these places become real to the potential traveller. Attachment can also be fostered through avatars, which are key constituents driving social interaction in VWs. They symbolise the self and therefore often represent our deepest wishes, aspirations, virtues, and vices (Bugeja, 2008) and can contribute to place identity. Avatars are also used to engage in social interactions, express emotions, opinions, and protest (Plunkett, 2011). Social interaction through gesture and avatar interactions can build identity (Ghani et al., 2019). This implies a strong person–avatar connection that leads to identity formation not only with the avatar but also the VE in which those interactions happen.

Perhaps the most researched area where sense of place has been shown to emerge from VR and AR experiences is in the field of heritage tourism. VR and AR can be used for heritage preservation (Guttentag, 2010; Liang & Elliot, 2020; Yung & Khoo-Lattimore, 2019). There is a recognition in the broader heritage experience literature that digital technology can support sense of place. This sense of place can emerge from the 'physical' attributes of the environment assessed through the user's site context, architecture, graphical fidelity, and soundscape (Ghani et al., 2019). Digital representations of cultural heritage sites, monuments, and buildings are commonly developed as interactive VEs, VR modelling language and high dynamic range imaging, or 3D walkthroughs (Ghani et al., 2019). In this virtual heritage environment sense of place has been linked to the 'presence' research domain (Ghani et al., 2019). As Malpas (2008) asserts, the loss of a sense of place and the object threatens a loss of sense of place of the subject, and with it, a loss of the proper sense of heritage. For example, virtual simulations of prehistoric landscapes are essentially forms of constructed space, which draws on aspects of the actual physical space of the present and the imagined space of the past. They can illuminate what is now imperceptible, contextualise what is now isolated and incongruous, and provide a means of connecting with people and cultures from which we are separated by thousands of years (Falconer, 2017). These experiences can deliver better learning and knowledge about historical and cultural heritage places among visitors (Ghani et al., 2019), and foster sense of place.

Within heritage environments (Falconer, 2017), there are several ways to use virtual technologies to create a sense of place for visitors. For example, Sequeira and Morgado (2013) identify four different ways in which VW platforms can be used for archaeology and heritage design experiences: First, environments can be created, including landscapes and communities to take on their own value over time and become 'historical' sites, imbued with their own cultural significance; second, reconstructive virtual archaeology, where physical world heritage sites can be simulated within the VW so that users can understand the physical configuration; third, virtual museums which replicate artefacts and/or visual images of real world artefacts; fourth, interactive virtual archaeology where users interact with the environment and each other by, for example, having avatars interacting with the reconstructed space (Falconer, 2017; Sequeira & Morgado, 2013). In VE, heritage-related meaning can be transferred from past memories, present memories, real place preferences, emotions, own culture, and heritage (Ghani et al., 2019), having implications for place identity and place dependence. Lin et al. (2020b) suggest that VR might lead to other forms of more sustainable travel such as slow travel for heritage destinations. Their findings indicate that VR is a very useful tool for encouraging respondents to travel to in a slower and more intensely observational manner, significantly arousing their sense of nostalgia and leading to a strong intention to actually slow travel to the destination in reality (Lin et al., 2020b). These examples highlight how VR and AR can be used to build a sense of place and also how to promote a destination or site to augment the reality of the experience or to immerse the visitor in a new and completely challenging tourism experience.

Conclusion

This chapter argues that sense of place can emerge from VEs through immersion, presence, flow, and social interactions, among others. Virtual spaces compete with actual places in providing meaning to users. In this way, the difference between reality and VR is blurring. The increasing number of users in VWs and the drastic changes in the external environment (e.g., COVID-19, economic recession, terrorist attacks, and disasters) are purportedly tilting human kind towards VWs where they can seemingly enjoy a safe, meaningful, and fulfilling experience, including tourism experience. As others have suggested, VR conveys a sense of escapism to the user (Beck et al., 2019; Guttentag, 2010). Tourism studies have focused primarily on the positive contributions of VR to tourism through tourist education, information provision, heritage management, gamification and destination marketing, among others (Beck et al., 2019; Guttentag, 2010; Tussyadiah et al., 2018; Yung & Khoo-Lattimore, 2019) and argued that virtual spaces enable users' enjoyment, build a sense of community, and the confidence of users (Laing et al., 2009). However, the negative facets of virtual spaces related to how they can exacerbate anxiety of users, increase ambiguity and confusion between 'real' and 'not real' (Laing et al., 2009), lead to cognitive overload and attention distraction (Suh & Prophet, 2018), as well as encourage loneliness and alienation (Laing et al., 2009) remain sparsely discussed in both the broader VR literature and tourism literatures. Whether VR tourism will replace actual travel and the authenticity of VR experiences remain hotly debated (Guttentag, 2010), with COVID-19 potentially further reigniting this debate. Nevertheless, the existing evidence suggests that immersion, presence, and flow are, for example, necessary but not sufficient for fostering a sense of place. This broad range of cognitive reactions must be accompanied by affective ones (e.g., enjoyment, fun and other positive/negative emotions) (Suh & Prophet, 2018) for place attachment to develop in VWs. Adaptation of both cognitive and affective attributes of VEs will be necessary for the optimal application of VR as a tourism tool (Yung & Khoo-Lattimore, 2019).

9 Conclusion

Home and away – losing and finding a sense of place

Introduction

We all have senses of place. As this book has shown, together with the experiences of the individual reader, a sense of place is one of the most important understandings of one's life providing a way not only to make environments legible but also to help feel secure. But as this book has shown, senses of place and place identities may not always be positive.

This final chapter posits some difficult questions about the nature of the relationship between place and tourism and the future of place in a rapidly changing world. This is an important issue as, if we are thinking in the context of the tourism system, then travels are contingent upon our understanding of place, and therefore our senses of place, in at least three, and possibly four, different ways. First, place in terms of home or at least usual or permanent residence. Second, place in terms of the destination of where one is going to travel to. Of course, in some cases this may even be 'home'. For example, as noted in the second home or summer house/cottage literature, in some cases the second home may even come to be regarded as being more permanent than the residential address as the holiday home is 'always there' and may be passed down through generations, providing a relative unchanging place of return to otherwise mobile family members (Aronsson, 2004; Hall, 2014b). Similarly, for migrants if the destination is the original origin country then return travel may also seem like a homecoming. Third, different senses of place are also attached to the range of potential destinations that an individual may select from. Each possible destination that is part of the particular choice set of destinations an individual selects from for travel purposes is also going to be attached to different understandings and senses of place. Finally, the fourth possible sense of place that applies in the tourism system is that of the route one selects to travel from place of residence to a destination and back. Clearly, sense of place here is also going to be connected to mode of travel but for those travelling by road and even by rail sense of place may influence the selection of transit routes, even in terms of choosing the most scenic options or travelling a particular way to rekindle a childhood, family, or other memory.

In all of the above three or four sense of place that are part of the tourism system, the tourism industry plays a part in influencing understandings of place as well as place sentiment. As so many tourism textbooks highlight (e.g. Cooper

& Hall, 2018), the motivations to travel are a result of both pull and push factors, while many of these motivations are no doubt authentic and genuine to the traveller they are also played (prayed?) upon by the tourism industry in their advertising and promotion seeking to persuade people that home and/or the place of normal residence is deficient in some way and that elsewhere lies the promise of happiness, satisfaction, self-discovery, love, and connection (Hall, 2011). Tourism therefore has an almost schizoid approach to sense of place because in its promotion it often seeks to market places in a way that seeks to simplify and influence understandings of place while often simultaneously seeking to claim authenticity. Similarly, while promoting a certain dimension of place tourism is almost certainly simultaneously changing the very nature of that place as a result of development (Sæþórsdóttir & Hall, 2020; Sæþórsdóttir, Hall, & Wendt, 2020). The underlying lesson from all this is that tourism is deeply embedded in changing places and senses of place. Indeed, change is emerging as one of the greatest challenges to sense of place and therefore the sense of wholeness and wellbeing that people can enjoy from positive senses of place.

While tourism *is* a major contributor to place change, it is only one of many contemporary processes such as globalisation, urbanisation, virtualisation, and global environmental change that are affecting place. This concluding chapter is going to briefly touch on several of these areas and highlight them as significant research challenges for tourism and sense of place in the future before returning to the theme of travel and tourism as a means of journeying and pilgrimage to find place.

Placing change, changing place

Places change together with people's relationships to place for a variety of reasons. As the various chapters in this book have highlighted, tourism- and migration-related mobility is a major factor in changing senses of place and place identity. For a growing minority of the world's population that almost certainly includes the readers of this book, there is an international labour and mobility market that encourages, if not necessitates, movement that is actively supported by countries, business, and institutions that seek to attract and retain talent and skills (Fink & Miguélez, 2017). This has meant that spatial and international mobility has become

> an exemplary characteristic of the global era and an important aspect of the cultural experience of many people in advanced industrialised nations. Mobility evokes powerful images that effectively counter any illusion of stationary life and provide a break from the insularity of the local and parochial. High levels of mobility are simultaneously a fact, a necessity and a cultural aspiration.
>
> (Skrbis, Woodward, & Bean, 2014, p. 614)

For those with university degrees, and especially for people with graduate qualifications, labour market mobility has become something of the norm, while this may be related to skill and language acquisition, it is also a part of developing international experience and working with international business (Amendola & Restaino, 2017; Carmona, Marhuenda Fluixá, Hernaiz-Agreda, & Navas Saurin, 2018). Universities and government-supported schemes even promote such mobility, which has acquired a status as a form of symbolic capital (Basaran & Olsson, 2018; Bauder, Hannan, & Lujan, 2017). However, this also means that people are moving more, creating multiple place attachments and connectivities. While at one level this may potentially mean that one becomes more cosmopolitan in outlook (Amit, 2015; Amit & Gardiner Barber, 2015), this may also have other implications for relationships to place and place identity as well as future mobility. Hall (2005), for example, argued that the more mobile one is in terms of visiting and living in places then the more mobile one would be in the future as, in order to retain social relationships with friends, family and significant others as well as maintaining overall place relations, it was necessary to return. From such a perspective while ICT such as email, Facebook, Skype, and Zoom are helpful in maintaining long-distance social, economic, and other relationships, Hall (2005) argued that they cannot as yet replace the importance of physical contact and conversation in place in sustaining relationships. Clearly, if such a perspective holds true then it has substantial implications for environmental and social futures in a travel and place context.

Maintaining multiple place attachments in an *active* sense therefore requires travel and in situ experience of place and the people and environment that one values in a place. Indeed, an important area for future research is the implications of active versus passive involvement in place relationships and their implications for attachment and travel behaviours, as well as their overlap with the maintenance of personal relationships. There is also substantial research that demonstrates that frequently moving residences can have significant negative effects on mental health and other health issues, particularly of younger people (Cuffee et al., 2014; Davey-Rothwell, German, & Latkin, 2008; Lowe & DeVerteuil, 2020; Raviv, Keinan, Abazon, & Raviv, 1990; Yuan & Manuel, 2018), with the stress being greater for long-distance relocations into new physical and cultural environments. In the case of people who are economically insecure then moving residence is part of a broader problem of housing security and homelessness that are also tied to the importance of maintaining community and place relations. However, an important question for the highly mobile is whether we are seeing another form of homelessness in the sense that having multiple place relations can make the notion of home, and the sense of security that it can provide, increasingly problematic? For example, for some people there is the pain of exile, knowing that you cannot return home because of political reasons or because home no longer exists as a result of physical and/or social destruction, e.g. because of deportation, displacement, forced migration or movement, genocide, pogroms, or massacres. When the Iranian singer and great interpreter of Persian music, Mohammad Reza Shajarian, died in October 2020, Brehmer (2020) observed,

> For Iranians abroad, Shajarian gave voice to the melancholy that comes with living in exile – not only physical exile, but also the existential inner exile

> of human beings, which is about living in separation from your innermost essence, a condition that Sufi poetry describes as the root of all human pain and suffering.

Exile is therefore a form of existential homelessness which may be temporary or, in some cases, become a permanent condition. While such a condition, together with wider feelings of estrangement and displacement, is clearly associated with refugees, migrants, and the members of diasporas (Ahmed, 1999; Brothers & Lewis, 2012; O'Neill & Spybey, 2003; see also Chapters 2 and 5), we should also ask to what extent are they also attached to the highly mobile and the expatriate. To what extant does travel and mobility lead people to lose a sense of home and place and become alienated and homeless? (D'Andrea, 2006; Labrador, 2002; Ralph, 2009). For example, Whitehead (2017) drew on 48 personal narratives professionals who had moved internationally more than once to examine how the notion of home is manifested as part of the life of the transnational professional. His findings revealed contradictory references to a sense of homelessness, which in the work context was perceived as ideal, but in the personal sphere was found to generate negative feelings of self-alienation. Although much of the world is still essentially sedentary, the growing mobility of some and the hypermobility of a minority does highlight how the notion of home as a single location with static roots is beginning to fall apart with corresponding implications for how we understand sense of place and tourism's role in the manufacturing of place.

A key research issue, and which was partly touched on in the opening chapters of this book, is how can place be constructed in a way in which feelings of home and senses of place are strengthened and reinforced for those who experience place. Cancellieri (2017, p. 49) writes that 'the ideas of placelessness and detachment from place … conceals the human necessity to constantly create new senses of home and new home-making practices'. He argues that people, in his study migrants, 'produce "home" by imbuing domestic spaces with their own memory and meaning and creating public and collective spaces characterised by "homely relations"' (Cancellieri, 2017, p. 49). For example, Zittoun (2020) comments:

> Families that move every couple of years for professional reasons, sometimes called 'expats', have complex and rich experiences of geographical mobility. Their children have often lived in five or six countries across two or more continents. These families develop strategies to maintain a stable sense of home, or family, across place. They invent family rituals, recreate the same living room with familiar pictures and ornaments, and some build a 'life in a box' for their children, containing the few toys and objects that follow them everywhere. It's similar for people in long trajectories of migration, such as asylum seekers arriving in Europe, who might have to sleep every night in a new shelter. Some build small objects to place around their bunks, to create a minimal home – an imagination of a stable place. Hence, when people's geographical mobility is too high and threatens their own sense of continuity, they use their imagination to create a more stable lifeworld.

As a result, home needs to be regarded not as a romanticised, fixed, and bounded place, but as 'a plural and conflictual field of action that can support social exclusion but can also open new interconnections and possibilities of peoples' empowerment' (Cancellieri, 2017, p. 49).

Undoubtedly, Cancellieri (2017) is correct in arguing that people produce home. However, at the same time we need to ask what are the material conditions that help produce place. A key theme of this book, and a concern of this last chapter with respect to its absence, is that even with increased mobility place(s) continues to be a basic part of the construction of a person and their relational self. As Hoey (2010, p. 237) observes,

> a sense of place, understood as manifest in personal attachment to real and imagined elements of particularity in place, which individuals understand as 'local character,' may support people in their ability to form lasting autobiographical accounts, expressions of individual 'character' critical to personhood.

In terms of how placemaking may be achieved, Arefi and Triantafillou (2005) argue that four ontological constructs of place need to be attended to in planning and design: as a set of visual attributes, product, process, and meaning (Table 9.1).

Importantly, the role of tourism in placemaking needs greater consideration (Lew, 2017) so that the placemaking process and the senses of place that are created become more meaningful for residents, workers, and visitors (Arefi, 2014). Therefore, a major research challenge is for tourism planning and design to examine the different ways in which senses of place are created and the economic and social implications of such place relationships understood.

An additional dimension of placemaking which is becoming increasingly important in the tourism context is that of the construction of virtual (online)

Table 9.1 Ontological constructs of place in design and planning

Ontology	*Related concepts and theories*
Place as a set of visual attributes	Image, scapes, picturesque, syntax, visual excitement, legibility, understanding, contradiction/complexity, iconography, visual layering, chaos
Place as product	Architectural typology and archetypes, new urbanism, morphology, experience economy, staged sets, city as theme park
Place as process	Uneven development, political economy of place, social, and cultural production of place
Place as meaning	Sense of place, sacred and the profane, mythical and real place, betweenness of place, phenomenology, territoriality, livability, non-places, virtual place, imaginary place

Source: After Arefi and Triantafillou (2005).

places and place relations. Virtualisation is the act of creating a virtual version of something and can both 'real' and 'imagined' places. Relph (2007) highlights that the mutual interaction between 'real' and virtual places will lead to people's experiences of real places being changed. As Chapter 8 emphasised, the experience of virtual tourism worlds and their contribution to senses of place will be profound. However, the implications of this remain unknown and are potentially evolving over time as the virtual world becomes more commonplace and ubiquitous. Undoubtedly, virtual place experiences can be invaluable for conveying a sense of 'pastness', as in highlighting what has been lost or can no longer be seen in terms of heritage (Holtorf, 2013). Devine (2017), for example, argues that 'time-based virtual heritage supports richer virtual heritage places, both phenomenologically and culturally, and that these in turn offer engaging, affective and memorable experiences creating learning opportunities for the general public'. However, we do not really understand the effects that this may have on accepting such changes to heritage or on how it influences travel behaviours with respect to the desire to engage in 'authentic' experiences. For example, the capacity to utilise augmented and virtual technologies in walking around places clearly affects the nature of engagement with place. As gardening author Monty Don has stated

> whenever he is in a city, "the thing I notice most – and what shocks me most – is how many people are walking along with headphones in, looking at a phone". He suggested, quite reasonably, that such sensory deprivation might be damaging. "They're not noticing anything at all – the weather, the sky, any other sounds. And that has to be bad for you".
>
> (quoted in Nicholson, 2020)

Therefore, while e-tourism and smart tourism is often touted by academics, destination managers, policy makers, and businesses as the future of tourism, there is insufficient consideration of what the implications of virtual and augmented technologies are for place and associated relationships.

As noted above, virtual technologies can act to substitute what has been lost at a destination – its 'pastness'. What may become an increasingly important consideration is the extent to which such technologies may serve to show not just long-lost heritage and landscapes but also these which have been lost in the very recent past or are disappearing in the present. Such coverage can include heritage buildings and locations (Seyfi & Hall, 2020), such as destroyed monuments or streetscapes, but can also include representations of landscapes and species lost to environmental change, particularly as the result of global heating. However, we do not know what the effects of such initiatives might be on how place is seen as well as the effects on behaviours. Chapter 7 described how strong senses of place can influence the adoption of pro-environmental behaviours. The impact of increased use of virtual and augmented technologies in place promotion and narratives on behaviours remains unknown.

Finally, with respect to the changing nature of places and hence sense of place, the impact of local and global environmental change will be profound on how places are perceived and what places are actually like. Feelings of solastalgia, the anxiety and despair that many people experience as their familiar places change as a result of disasters, such as earthquakes (see Chapter 6) or COVID-19, or environmental changes, such as fire, drought, storm events and winds, and sea-level rise, and erosion are only likely to increase. The extent to which place change will influence feelings of helplessness and overall wellbeing is unknown, but the experiences of countries that have had extended droughts or fire events, such as Australia, would suggest that the impact on community health, including sense of place, will be major (Cianconi, Betrò & Janiri, 2020; Ellis & Albrecht, 2017; Horton, Hanna, & Kelly, 2010). Therefore, there is an urgent need to find ways not only to prevent global environmental change but also to develop adaptive strategies that may minimise the 'homesickness' that can occur when the place(s) that you call home changes and becomes unrecognisable.

Box 9.1 Humanistic pilgrimage and the search for meaning

Humanists can engage in pilgrimage in which meaning is intended to be derived from both the process of travel and the object, or destination of such travel (Hall, 2006a). Elements of such an approach towards pilgrimage can be witnessed in the travel of people to find their family history or significant places, people, and events in their lives. With respect to family history the desire to find out from whence one came is a response to issues of identity and place (Coles & Timothy, 2004).

The location of meaning in secular pilgrimage can be understood in terms of travel to different places and locations. For example, for Australians, New Zealanders and Turkish people travel to Gallipoli Peninsula to witness the site of the 1915 invasion is a secular pilgrimage of national identity in which the myths of nationhood and cultural identity are paramount (Hall, 2002). Similarly, meaning in secular pilgrimage can also be fulfilled by other places, such as sporting grounds (e.g., for many Australians the Melbourne Cricket Ground is often described as a 'holy of holies' because of its association with Australian Rules Football and Cricket) and even with respect to popular entertainment, such as Elvis Presley (Doss, 1999; Rinschede & Bhardwaj, 1990), the Beatles (Kruse, 2003), or even television programmes such as *Star Trek* (Jindra, 1994; Porter & McLaren, 1999; Porter, 2004). Importantly, the subjects of such secular pilgrimages are also place-makers, for example, it is almost impossible to think of Liverpool without also thinking of the Beatles (Kruse, 2005).

To some people, regarding a visit to Graceland (Elvis Presley), Bertels Salon Kompagnistæde (cheesecake), Postgatan Kalmar (tapas), Cardiff waterfront (*Torchwood*), Doune Castle (*Monty Python and the Holy Grail*), Stamford Bridge (Chelsea's football stadium), or Strawberry Fields (Beatles) as pilgrimage, and therefore placing such places in a category with that of Canterbury, Jerusalem, Mecca, or Rome may seem like blasphemy. But why should it? If one takes the religious humanist position that humans have created god then if a visit to Liverpool has a deep personal meaning that helps people explain their lives then it is of equivalent value to that of a Christian to Jerusalem, or a Muslim to Mecca. As Kruse (2003) observes in locating Strawberry Fields as a place of pilgrimage, in world religions such as Buddhism, Christianity, or Islam, pilgrimage is the physical/exteriorised quest for the holy.

A pilgrim leaves home and the familiar world behind on a journey to a sacred centre(s) 'out there' often through extremely proscribed routes and forms (Campo, 1991). But, more importantly, pilgrimage is also a journey 'in there', a spiritual interior quest within the heart of those who feel something missing in their lives – a sense of mystery and wonder, wellbeing, meaning, and connection with others (Coleman & Elsner, 1995; Turner, 1973). Indeed, as the Very Reverend Robert Willis, Dean of Canterbury observed, with respect to the search for 'blessing and enrichment. In pilgrimage, body, mind and spirit come together in an individual quest… Jesus was always walking, walking, walking – all the way to Calvary'. The quest also need not be momentous: "Any journey that adds a mini-jigsaw piece to the puzzle of you can be a mini-pilgrimage' (quoted in Chu, 2004). Similarly, Iyer (1999, p. vii) commented, 'A pilgrim does not have to be moving towards something holy… so much as toward whatever resides in the deepest part of him: it could be a poet who gave wings to his soul, or a lover who broke his heart open'. The journey and the place therefore go hand-in-hand.

The overt search for meaning by religious humanists may even occupy the same pilgrimage spaces as those from more traditional theistic faiths. For example, the pilgrimage walk to Santiago de Compostela on the Camino de Santiago or the Pilgrim's Way in Kent to Canterbury is undertaken by believers and non-believers alike, but the journey is important to all. The liminality and communitas of pilgrimage is not the sole domain of the religious. For example, Chu (2004) notes that 'Part of the joy of pilgrimage is a spirit of community that comes from identifying with something bigger than oneself. The pilgrim who sets out solo shares a bond with others who journey on the same path: the aches, the pains and the triumphs' Yet, as Chu (2004) went on to note 'each pilgrim is utterly alone, because a pilgrimage is a trip not just to a physical place but also into a person's soul'. Such a theme was picked up in Bouldrey's (1999) collection of essays and short stories of contemporary pilgrimage entitled *Travelling Souls*. The majority of stories are observations of the spiritual dimensions of secular pilgrimage and quest. Someone who goes on a pilgrimage is usually quite clear about the route and the destination. A pilgrimage is, in one sense, a very regular sort of journey, often taken amidst a big crowd of other people all with their eyes on the same goal. As Bouldrey (1999,

p. xvi) comments, 'Every person might take the same pilgrimage and bring home another story'. By contrast, someone who is going on a quest may know what they are looking for, but they do not know the route. Nevertheless, the journey is still worthwhile as it is the journey in which one's heart remains open 'to little birds' (ee cummings in Reef, 2006, p. 95) and the stopping places where one rests, if even for just a short time on that journey wherever in the world that are actually what are most important (Hall, 2006a).

Place, journeying, pilgrimage, and the search for home

Sense of place, place attachment, and notions of home are central to travel and tourism. However, in looking at many modern tourism textbooks it often seems that there is a presentation of 'tourism as modernity' versus 'pilgrimage as pre-modernity' as if part of some seamless evolutionary development that has undergone a gradual and 'normal' transition. Yet, the history of tourism is often written from particular secular, consumerist, and limited timescale perspectives and fails to acknowledge the enormous diversity of travel motivations and experiences that occupy and sometimes compete within the same space and time (Hall, 2006a; Hall & Prayag, 2020).

As noted in many of the chapters of this book, and Chapter 6 in particular, even in this so-called postmodern period, religious pilgrimage is alive and well, even though much of the contemplation of tourism movement is undertaken in a secular fashion. There is therefore, arguably, a disjoint between much of the actual nature of tourism and human mobility as is experienced by many of the people undertaking it and what is often being described in terms of voluntary mobility, or what we usually describe as tourism, at various scales. Much of this disconnect probably occurs because much discussion of tourism tends to focus on what happens at destinations or, to be more precise, at a specific point of time at a destination, rather than looking at the entire tourism system and the connection between home and away or, to put it another way, as a connection between places in which what happens in one place can only be more completely explained if you also understanding the relationship of the tourist to the other.

Furthermore, there are insufficient studies of tourism that seek to explain tourist mobility within the context of the overall lifecourses of individuals (Hall, 2005), although we begin to gain glimpses of explanation when we engage in research on migration, exile, and the influences of early travel experiences on later behaviours and consumption choices. In other words, there is a need to provide a much more holistic view of people's mobility that is just as interested in the actions, behaviours, and influences of people at 'home' as it is when they are away. This therefore means trying to avoid placing people in the overly simplistic category of 'tourist' and just focusing on what people do away from the home environment. We cannot understand place relationships and future travel behaviour unless we also understand the range of influences on an individual's path through life and their relationship to the different places they experience.

Therefore, the study of sense of place and place attachment in a tourism context highlights the need for a more gestalt perspective of tourism and traveling which also suggests the need to examine what happens in the spiritual life and experiences of people in their home environment as this must also inform their travel behaviours, even if sometimes in unconscious ways. From such a perspective, notions of religion or the sacred and the secular, or spirituality and the secular, should not be seen as opposites and are instead highly permeable and intermeshed concepts.

The idea of journeying is therefore extremely important as it implies travel from one place to another. Furthermore, as Iyer (1999, p. vii) observes,

> every journey is a question of sorts, and the best journeys for me are the ones in which every answer opens onto deeper and more searching questions. Every traveler is on a quest of sorts, but the pilgrim stands out because his every step is a leap of faith.

However, journeying may take several forms. Ideally, the journey makes us more experienced people if not better people as it is a process of learning, of social relations, and conscious connections with places that serves to encourage reflection on life, our values, and our ethics. Indeed, the very notion of pilgrimage is apt for those who leave the security of traditional mono-theistic religions that have a God independent of man as the word 'pilgrim' comes from Latin, per agrum: one who leaves the security of village and road and walks 'through the fields' (Cheer, 2000). This is not to say that all travel is continually consciously seen as a spiritual pilgrimage. Even the most ardent religious pilgrims have difficulty in concentrating on the spiritual rather than the secular and the banal over the entire course of their journey while different pilgrims are also seeking different things from the same route or sacred space.

To engage with place and the significance of its meaning and its role in identity and attachment is to become aware that all life is a journey in which we are 'essentially traveling deeper into faith and doubt at the same time: deeper into complexity' (Iyer, 1999, pp. viii–ix). No step on a trip or place you stop is ever wasted. Being aware of the nature and importance of place, and the relationality between places and tourists is therefore integral to the development of a more conscious and aware tourism and traveller. Such journeys and engagement with places can be undertaken on foot and sometime vicariously, they are invariably taken, to varying degrees, with others – for which we should always be grateful. Yet the important thing is that the journey is taken.

References

Ablett, P. G., & Dyer, P. K. (2009). Heritage and hermeneutics: Towards a broader interpretation of interpretation. *Current Issues in Tourism*, 12(3), 209–233.

Adie, B. A. (2020). Place attachment and post-disaster decision-making in a second home context: A conceptual framework. *Current Issues in Tourism*, 23(10), 1205–1215.

Ahmed, S. (1999). Home and away: Narratives of migration and estrangement. *International Journal of Cultural Studies*, 2(3), 329–347.

Aitken, R., & Campelo, A. (2011). The four Rs of place branding. *Journal of Marketing Management*, 27(9–10), 913–933.

Albrecht, G. (2005). 'Solastalgia'. A new concept in health and identity. *PAN: Philosophy Activism Nature*, 3, 41–55.

Alkon, A. H., & Traugot, M. (2008). Place matters, but how? Rural identity, environmental decision making, and the social construction of place. *City & Community*, 7(2), 97–112.

Allison, E. (2019). Deity citadels: Sacred sites of bio-cultural resistance and resilience in Bhutan. *Religions*, 10(4), 268–285.

Allport, G. W. (1966). The religious context of prejudice. *Journal for the Scientific Study of Religion*, 5(3), 447–457.

Alston, M., Hargreaves, D., & Hazeleger, T. (2018). Postdisaster social work: Reflections on the nature of place and loss. *Australian Social Work*, 71(4), 405–416.

Amaro, S., Antunes, A., & Henriques, C. (2018). A closer look at Santiago de Compostela's pilgrims through the lens of motivations. *Tourism Management*, 64, 271–280.

Amendola, A., & Restaino, M. (2017). An evaluation study on students' international mobility experience. *Quality & Quantity*, 51(2), 525–544.

Amit, V. (2015). Circumscribed cosmopolitanism: Travel aspirations and experiences. *Identities*, 22(5), 551–568.

Amit, V., & Gardiner Barber, P. (2015). Mobility and cosmopolitanism: Complicating the interaction between aspiration and practice. *Identities*, 22(5), 543–550.

Amoamo, M. (2011). Tourism and hybridity: Re-visiting Bhabha's third space. *Annals of Tourism Research*, 38(4), 1254–1273.

Amoamo, M., & Thompson, A. (2010). (re) Imaging Māori tourism: Representation and cultural hybridity in postcolonial New Zealand. *Tourist Studies*, 10(1), 35–55.

Amoamo, M., Ruckstuhl, K., & Ruwhiu, D. (2018). Balancing indigenous values through diverse economies: A case study of Māori ecotourism. *Tourism Planning & Development*, 15(5), 478–495.

Amore, A., & Hall, C. M. (2016). From governance to meta-governance in tourism? Re-incorporating politics, interests and values in the analysis of tourism governance. *Tourism Recreation Research*, 41(2), 109–122.

Amore, A., & Hall, C. M. (2017). National and urban public policy in tourism. Towards the emergence of a hyper-neoliberal script? *International Journal of Tourism Policy*, 7(1), 4–22.

Anant, S. S. (1966). Need to belong. *Canada's Mental Health*, 14(2), 21–27.

Andriotis, K., & Vaughan, R. D. (2003). Urban residents' attitudes toward tourism development: The case of Crete. *Journal of Travel Research*, 42(2), 172–185.

Animesh, A., Pinsonneault, A., Yang, S. B., & Oh, W. (2011). An odyssey into virtual worlds: Exploring the impacts of technological and spatial environments on intention to purchase virtual products. *MIS Quarterly*, 35(3), 789–810.

Aparna, K., & Schapendonk, J. (2018). Shifting itineraries of asylum hospitality: Towards a process geographical approach of guest-host relations. *Geoforum*. https://doi.org/10.1016/j.geoforum.2018.03.024

Archibald, J. A. (2008). *Indigenous storywork: Educating the heart, mind, body, and spirit*. Vancouver, BC: University of British Columbia Press.

Arefi, M., & Triantafillou, M. (2005). Reflections on the pedagogy of place in planning and urban design. *Journal of Planning Education and Research*, 25(1), 75–88.

Arefi, M. (2014). *Deconstructing placemaking: Needs, opportunities, and assets*. Abingdon: Routledge.

Arnould, E. J., & Price, L. L. (1993). River magic: Extraordinary experience and the extended service encounter. *Journal of Consumer Research*, 20(1), 24–45.

Aronsson, L. (2004). Place attachment of vacation residents: Between tourists and permanent residents. In C. M. Hall & D. K. Müller (Eds.), *Tourism, mobility and second homes: Between elite landscape and common ground* (pp. 75–86). Clevedon, UK: Channelview.

Arruda, W. (2003). *Urban voice – could Santa be the world's strongest brand?* http://www.brandchannel.com/brand_speak.asp?bs_id=74

Ashforth, B. E., Harrison, S. H., & Corley, K. G. (2008). Identification in organizations: An examination of four fundamental questions. *Journal of Management*, 34(3), 325–374.

Ateljevic, I. (2000). Circuits of tourism: Stepping beyond the 'production/consumption' dichotomy. *Tourism Geographies*, 2(4), 369–388.

Åkerlund, U., Lipkina, O., & Hall, C. M. (2015). Second home governance in the EU: In and out of Finland and Malta. *Journal of Policy Research in Tourism, Leisure and Events*, 7(1), 77–97.

Balaz, V. (1995). Five years of economic transition in Slovak tourism: Successes and shortcomings. *Tourism Management*, 16(2), 143–150.

Baños, R. M., Botella, C., Alcañiz, M., Liaño, V., Guerrero, B., & Rey, B. (2004). Immersion and emotion: Their impact on the sense of presence. *Cyberpsychology & Behavior*, 7(6), 734–741.

Barrios, R. (2011). "If you did not grow up here, you cannot appreciate living here": Neoliberalism, space-time, and affect in Post-Katrina recovery planning. *Human Organization*, 70(2), 118–127.

Basaran, T., & Olsson, C. (2018). Becoming international: On symbolic capital, conversion and privilege. *Millennium*, 46(2), 96–118.

Bassano, C., Barile, S., Piciocchi, P., Spohrer, J. C., Iandolo, F., & Fisk, R. (2019). Storytelling about places: Tourism marketing in the digital age. *Cities*, 87, 10–20.

Basu, P. (2004). My own island home: The Orkney homecoming. *Journal of Material Culture*, 9(1), 27–42.

Bauböck, R. (1994). *Transnational citizenship: Membership and rights in international migration*. Aldershot, UK: Edward Elgar.

Bauder, H., Hannan, C. A., & Lujan, O. (2017). International experience in the academic field: Knowledge production, symbolic capital, and mobility fetishism. *Population, Space and Place*, 23(6), e2040.

Baumeister, R. F., & Newman, L. S. (1994). How stories make sense of personal experiences: Motives that shape autobiographical narratives. *Personality and Social Psychology Bulletin*, 20(6), 676–690.

BBC (2002). *What do pagans believe in?* Retrieved from https://www.bbc.co.uk/religion/religions/paganism/beliefs/beliefs.shtml

Beck, J., Rainoldi, M., & Egger, R. (2019). Virtual reality in tourism: A state-of-the-art review. *Tourism Review*, 74(3), 586–612.

Beech, S. E. (2014). Why place matters: Imaginative geography and international student mobility. *Area*, 46(2), 170–177.

Beeton, S. (2016). *Film-induced tourism* (2nd ed.). Bristol, UK: Channel view publications.

Belhassen, Y., Caton, K., & Stewart, W. P. (2008). The search for authenticity in the pilgrim experience. *Annals of Tourism Research*, 35(3), 668–689.

Benford, S., Bowers, J., Fahlen, L. E., Mariani, J., & Rodden, T. (1994). Supporting cooperative work in virtual environments. *The Computer Journal*, 37(8), 653–668.

Berg, N. G. (2020). Geographies of wellbeing and place attachment: Revisiting urban–rural migrants. *Journal of Rural Studies*, 78, 438–446.

Berglund, E., & Olsson, K. (2010). *Rethinking place marketing – A literature review.* Paper presented in *Special Session: Planning and Place Marketing – theoretical implications, 50Th ERSA Congress*, Jönköping, Sweden, 19–23 August.

Besio, K. A. (2012). *Three kingdoms and Chinese culture*. Albany, NY: State University of New York Press.

Bhuian, S. N., Sharma, S. K., Butt, I., & Ahmed, Z. U. (2018). Antecedents and pro-environmental consumer behavior (PECB): The moderating role of religiosity. *Journal of Consumer Marketing*, 35(3), 287–299.

Bitner, M. J. (1992). Servicescapes: The impact of physical surroundings on customers and employees. *Journal of Marketing*, 56(2), 57–71.

Blizard, C. R., & Schuster, R. M. (2007). Fostering children's connections to natural places through cultural and natural history storytelling. *Children Youth and Environments*, 17(4), 171–206.

Bonaiuto, M., Fornara, F., & Bonnes, M. (2003). Indexes of perceived residential environment quality and neighbourhood attachment in urban environments: A confirmation study on the city of Rome. *Landscape and Urban Planning*, 65(1–2), 41–52.

Bonaiuto, M., Alves, S., De Dominicis, S., & Petruccelli, I. (2016). Place attachment and natural hazard risk: Research review and agenda. *Journal of Environmental Psychology*, 48, 33–53.

Booms, B. H., & Bitner, M. J. (1981). Marketing strategies and organization structures for service firms. In J. Donnelly & W. R. George (Eds.), *Marketing of services* (pp. 47–52). Chicago, IL: American Marketing Association.

Boon, H. J. (2014). Disaster resilience in a flood-impacted rural Australian town. *Natural Hazards*, 71(1), 683–701.

Boswell, R. (2005). Unravelling le malaise Creole: Hybridity and marginalisation in Mauritius. *Identities: Global Studies in Culture and Power*, 12(2), 195–221.

Bouldrey, B. (Ed.) (1999). *Travelling souls: Contemporary pilgrimage stories*. San Francisco, CA: Whereabouts Press.

Boulle, P. (1963). *La planète des singes [Planet of the Apes]*. Paris, France: René Julliard.

Bowen, J. R. (2002). *Religions in practice: An approach to the anthropology of religion*. Boston, MA: Allyn & Bacon.

Brandth, B., Haugen, M. S., & Kramvig, B. (2010). "The future can only be imagined"– Innovation in farm tourism from a phenomenological perspective. *Open Social Science Journal*, 3(1), 51–59.

Breakwell, G. (1983). *Threatened identities*. Chichester, UK: Wiley.

Breakwell, G. (1986). *Coping with threatened identities*. London, UK: Methuen.

Breek, P., Hermes, J., Eshuis, J., & Mommaas, H. (2018). The role of social media in collective processes of place making: A study of two neighborhood blogs in Amsterdam. *City & Community*, 17(3), 906–924.

Brehmer, M. (2020). Mohammad Reza Shajarian embodied the timeless beauty of Persian music. *The Guardian*, 9 October, https://www.theguardian.com/music/2020/oct/09/mohammad-reza-shajarian-iran-persian-music

Britton, S. (1991). Tourism, capital, and place: Towards a critical geography of tourism. *Environment and Planning D: Society and Space*, 9(4), 451–478.

Brockhall, F. J., & Liu, S. (2011). Performing new Australians: Identity (re) construction of long-term Greek and Cypriot immigrants in Australia. *China Media Research*, 7(1), 16–24.

Broomhall, S., & Spinks, J. (2010). Interpreting place and past in narratives of Dutch heritage tourism. *Rethinking History*, 14(2), 267–285.

Brothers, D., & Lewis, J. (2012). Homesickness, exile, and the self-psychological language of homecoming. *International Journal of Psychoanalytic Self Psychology*, 7(2), 180–195.

Brown, B. B., Perkins, D. D., & Brown, G. (2003). Place attachment in a revitalizing neighborhood: Individual and block levels of analysis. *Journal of Environmental Psychology*, 23(3), 259–271.

Brown, B. B., & Werner, C. M. (1985). Social cohesiveness, territoriality, and holiday decorations: The influence of cul-de-sacs. *Environment and Behavior*, 17(5), 539–565.

Brown, D. (2003). *The da Vinci code*. Sydney, Australia: Random House.

Bryon, J. (2012). Tour guides as storytellers–from selling to sharing. *Scandinavian Journal of Hospitality and Tourism*, 12(1), 27–43.

Bryson, S. (1995). Approaches to the successful design and implementation of VR applications. In A. E. Rae, A. V. John, & H. Jones (Eds.), *Virtual reality applications* (pp. 3–15). London: Academic Press.

Buchmann, A. (2006). From Erewhon to Edoras: Tourism and myths in New Zealand. *Tourism Culture & Communication*, 6(3), 181–189.

Buckley, R., Shekari, F., Mohammadi, Z., Azizi, F., & Ziaee, M. (2020). World Heritage tourism triggers urban–rural reverse migration and social change. *Journal of Travel Research*, 59(3), 559–572.

Bugeja, M. J. (2008). Second thoughts about second life. *The Education Digest*, 73(5), 18–22.

Burgen, S. (2017). Barcelona anti-tourism activists vandalise bikes and bus. *The Guardian*, 1 August. Retrieved from https://www.theguardian.com/world/2017/aug/01/barcelona-anti-tourism-activists-vandalise-bikes-and-bus

Burton, N. (2020). The coronavirus exposes the history of racism and "cleanliness". *Vox*, 7 February, https://www.vox.com/2020/2/7/21126758/coronavirus-xenophobia-racism-china-asians.

Butts, D., & Adams, H. (2020). Weather contracts: Capturing a sense of weather for place-based adaptation to climate change. *Global Environmental Change*, 63. https://doi.org/10.1016/j.gloenvcha.2020.102052

Buzinde, C. N., Kalavar, J. M., Kohli, N., & Manuel-Navarrete, D. (2014). Emic understandings of Kumbh Mela pilgrimage experiences. *Annals of Tourism Research*, 49, 1–18.

Caidi, N. (2019). Pilgrimage to Hajj: An information journey. *The International Journal of Information, Diversity, & Inclusion*, 3(1). https://doi.org/10.33137/ijidi.v3i1.32267

Campbell, J. (1991). *The power of myth*. New Jersey, NJ: Anchor.

Campbell, N. (2005). Producing America: Redefining post-tourism in the global media age. In D. Crouch, R. Jackson, & F. Thompson (Eds.), *The media and the tourist imagination: Converging cultures* (pp. 198–214). London, UK: Routledge.

Campo, J. E. (1991). Authority, ritual, and spatial order in Islam: The pilgrimage to Mecca. *Journal of Ritual Studies*, 5(1), 65–91.

Cancellieri, A. (2017). Towards a progressive home-making: The ambivalence of migrants' experience in a multicultural condominium. *Journal of Housing and the Built Environment*, 32(1), 49–61.

Carmona, C., Marhuenda Fluixá, F., Hernaiz-Agreda, N., & Navas Saurin, A. A. (2018). Educated for migration? Blind spots around labor market conditions, competence building, and international mobility. *European Educational Research Journal*, 17(6), 809–824.

Carpino, O. A., Berg, A. A., Quinton, W. L., & Adams, J. R. (2018). Climate change and permafrost thaw-induced boreal forest loss in northwestern Canada. *Environmental Research Letters*, 13(8), 084018.

Carr, A. (2017). Maori tourism in New Zealand. In M. Whitford, L. Ruhanen, & A. Carr (Eds.), *Indigenous tourism cases from Australia and New Zealand* (pp. 145–162). Oxford, UK: Goodfellow.

Carter, J., Dyer, P., & Sharma, B. (2007). Dis-placed voices: Sense of place and place-identity on the Sunshine Coast. *Social & Cultural Geography*, 8(5), 755–773.

Casey, E. S. (2001). Between geography and philosophy: What does it mean to be in the place-world? *Annals of the Association of American Geographers*, 91(4), 683–693.

Castles, S., & Davidson, A. (2000). *Citizenship and migration: Globalization and the politics of belonging*. New York, NY: Psychology Press.

Cave, D. (2012). *Digital islands: How the Pacific's ICT revolution is transforming the region*. Sydney: Lowy Institute for International Policy.

Cave, J., & Hall, C. M. (2015). Do families hold the Pacific together? VFR, voyaging and new expressions of diasporic networks. In E. Backer & B. King (Eds.), *VFR travel research: International perspectives* (pp. 187–204). Bristol, UK: Channel View.

Chakravarti, A. K. (1974). Regional preference for food: Some aspects of food habit patterns in India. *Canadian Geographer/Le Géographe Canadien*, 18(4), 395–410.

Chamlee-Wright, E., & Storr, V. H. (2009). "There's no place like New Orleans": Sense of place and community recovery in the Ninth Ward after Hurricane Katrina. *Journal of Urban Affairs*, 31(5), 615–634.

Chan, J. H., Chen, S. Y., Ji, Z., Zhang, Y., & Qi, X. (2020). Cases of indigenous entrepreneurship in mainland China and Taiwan. In R. Colbourne & R. B. Anderson (Eds.), *Indigenous wellbeing and enterprise: Self-determination and sustainable economic development*. New York, NY: Routledge.

Chan, K. W. (2009). The Chinese hukou system at 50. *Eurasian Geography and Economics*, 50(2), 197–221.

Chan, K. W., & Zhang, L. (1999). The hukou system and rural-urban migration in China: Processes and changes. *The China Quarterly*, 160, 818–855.

Chang, R. C., Kivela, J., & Mak, A. H. (2010). Food preferences of Chinese tourists. *Annals of Tourism Research*, 37(4), 989–1011.

Chang, Y. H. (2012). *Food and identity: A socio-historical perspective on the evolution of Taiwanese cuisine* (Doctoral thesis). Griffith University: Griffith University Research Repository. [pdf]. Retrieved from https://research-repository.griffith.edu.au/bitstream/handle/10072/367854/Chang_2012_02Thesis.pdf?sequence=1

Chapin III, F. S., & Knapp, C. N. (2015). Sense of place: A process for identifying and negotiating potentially contested visions of sustainability. *Environmental Science & Policy*, 53, 38–46.

Cheer, J., & Reeves, K. (2013). Roots tourism: Blackbirding and the South Sea Islander diaspora. *Tourism Analysis*, 18(3), 245–257.

Cheer, J. M., Belhassen, Y., & Kujawa, J. (2017). The search for spirituality in tourism: Toward a conceptual framework for spiritual tourism. *Tourism Management Perspectives*, 24, 252–256.

Cheer, N. (2000). Pilgrimage: Linear and circular. *Sea of Faith Network N.Z., SOF Newsletter*, 36, May.

Chen, N. C., & Dwyer, L. (2018). Residents' place satisfaction and place attachment on destination brand-building behaviors: Conceptual and empirical differentiation. *Journal of Travel Research*, 57(8), 1026–1041.

Chen, N. C., Dwyer, L., & Firth, T. (2014a). Conceptualization and measurement of dimensionality of place attachment. *Tourism Analysis*, 19(3), 323–338.

Chen, N. C., Dwyer, L., & Firth, T. (2014b). Effect of dimensions of place attachment on residents' word-of-mouth behavior. *Tourism Geographies*, 16(5), 826–843.

Chen, N. C., Dwyer, L., & Firth, T. (2015). Factors influencing Chinese students' behavior in promoting Australia as a destination for Chinese outbound travel. *Journal of Travel & Tourism Marketing*, 32(4), 366–381.

Chen, N. C., Dwyer, L., & Firth, T. (2018). Residents' place attachment and word-of-mouth behaviours: A tale of two cities. *Journal of Hospitality and Tourism Management*, 36, 1–11.

Chen, N. C., Qi, S., & Hall, C. M. (2020). Halal food certification in China. In C. M. Hall & G. Prayag (Eds.), *The Routledge handbook of halal hospitality and Islamic tourism* (pp. 249–263). Abingdon, UK: Routledge.

Chen, N. C., Hall, C. M., Yu, K., Qian, C. (2019). Environmental satisfaction, residential satisfaction, and place attachment: The cases of long-term residents in rural and urban areas in China. *Sustainability*, 11, 6439. doi:10.3390/su11226439

Chen, S., Wang, S., & Xu, H. (2017). Influence of place identity on residents' attitudes to dark tourism. *Journal of China Tourism Research*, 13(4), 338–356.

Cheng, T., & Selden, M. (1994). The origins and consequences of China's Hukou's System. *The China Quarterly*, 139, 644–668.

Cheng, T. M., Wu, H. C., & Huang, L. M. (2013). The influence of place attachment on the relationship between destination attractiveness and environmentally responsible behavior for island tourism in Penghu, Taiwan. *Journal of Sustainable Tourism*, 21(8), 1166–1187.

Cheong, R. (1995). The virtual threat to travel and tourism. *Tourism Management*, 16(6), 417–422.

Choe, J., & O'Regan, M. (2020). Faith manifest: Spiritual and mindfulness tourism in Chiang Mai, Thailand. *Religions*, 11(4), 177. https://doi.org/10.3390/rel11040177

Choi, H. S. C., & Sirakaya, E. (2005). Measuring residents' attitude toward sustainable tourism: Development of sustainable tourism attitude scale. *Journal of Travel Research*, 43(4), 380–394.

Christofi, V., & Thompson, C. L. (2007). You cannot go home again: A phenomenological investigation of returning to the sojourn country after studying abroad. *Journal of Counseling & Development*, 85, 53–63.

Chronis, A. (2005). Coconstructing heritage at the Gettysburg storyscape. *Annals of Tourism Research*, 32(2), 386–406.

Chronis, A. (2008). Co-constructing the narrative experience: Staging and consuming the American Civil War at Gettysburg. *Journal of Marketing Management*, 24(1–2), 5–27.

Chronis, A. (2012a). Tourists as story-builders: Narrative construction at a heritage museum. *Journal of Travel & Tourism Marketing*, 29(5), 444–459.

Chronis, A. (2012b). Between place and story: Gettysburg as tourism imaginary. *Annals of Tourism Research*, 39(4), 1797–1816.

Chu, J. (2004), The roads now taken. *Time Europe*, 164(1) July 5.

Chuang, F. M., Morgan, R. E., & Robson, M. J. (2012). Clan culture, strategic orientation and new product performance in Chinese marketing ventures: An exploration of main and moderating effects. *Journal of Strategic Marketing*, 20(3), 267–286.

Cianconi, P., Betrò, S., & Janiri, L. (2020). The impact of climate change on mental health: A systematic descriptive review. *Frontiers in Psychiatry*, 11, 74.

Clark, A. M., & Clark, M. T. (2016). *Pokémon Go and research: Qualitative, mixed methods research, and the super complexity of interventions*. Los Angeles, CA: Sage.

Clark, W. A., & Lisowski, W. (2019). Extending the human capital model of migration: The role of risk, place, and social capital in the migration decision. *Population, Space and Place*, 25(4), e2225.

Clarke, D., Murphy, C., & Lorenzoni, I. (2016). Barriers to transformative adaptation: Responses to flood risk in Ireland. *Journal of Extreme Events*, 3(2). https://doi.org/10.1142/S234573761650010X

Clarke, D., Murphy, C., & Lorenzoni, I. (2018). Place attachment, disruption and transformative adaptation. *Journal of Environmental Psychology*, 55, 81–89.

Cloke, P., & Conradson, D. (2018). Transitional organisations, affective atmospheres and new forms of being-in-common: Post-disaster recovery in Christchurch, New Zealand. *Transactions of the Institute of British Geographers*, 43(3), 360–376.

Coe, K., Aiken, N. E., & Palmer, C. T. (2006). Once upon a time: Ancestors and the evolutionary significance of stories. *Anthropological Forum*, 16(1), 21–40.

Coghlan, A., McLennan, C. L., & Moyle, B. (2017). Contested images, place meaning and potential tourists' responses to an iconic nature-based attraction 'at risk': The case of the Great Barrier Reef. *Tourism Recreation Research*, 42(3), 299–315.

Cohen, E., & Avieli, N. (2004). Food in tourism: Attraction and impediment. *Annals of Tourism Research*, 31(4), 755–778.

Cohen, R. (1997). *Global diasporas*. London, UK: Routledge.

Cohen, S. A., Duncan, T., & Thulemark, M. (2015). Lifestyle mobilities: The crossroads of travel, leisure and migration. *Mobilities*, 10(1), 155–172.

Coleman, S., & Elsner, J. (1995). *Pilgrimage past and present: Sacred travel and sacred space in the world religions*. Cambridge, MA: Harvard University Press.

Coles, T. (2008). Citizenship and the state: Hidden features in the internationalisation of tourism. In T. Coles & C. M. Hall (Eds.), *International business and tourism: Global issues, contemporary interactions* (pp. 55–69). London, UK: Routledge.

Coles, T., & Hall, C. M. (2006). The geography of tourism is dead. Long live geographies of tourism and mobility. *Current Issues in Tourism*, 9(4–5), 289–292.

Coles, T., & Hall, C. M. (2011). Rights and regulation of travel and tourism mobility. *Journal of Policy Research in Tourism, Leisure and Events*, 3(3), 209–223.

Coles, T., & Timothy, D. J. (Eds.). (2004). *Tourism, diasporas and space*. London, UK: Routledge.

Collins-Kreiner, N. (2010). Researching pilgrimage: Continuity and transformations. *Annals of Tourism Research*, 37(2), 440–456.

Collins-Kreiner, N. (2020). Religion and tourism: A diverse and fragmented field in need of a holistic agenda. *Annals of Tourism Research*, 82. https://doi.org/10.1016/j.annals.2020.102892

Connell, J. (2010). Pacific islands in the global economy: Paradoxes of migration and culture. *Singapore Journal of Tropical Geography*, 31, 115–129.

Connell, J., & Brown, R. P. C. (2005). *Remittances in the Pacific: An overview*. Manila: Asian Development Bank.

Consulate of Spain. (n.d.). *Booking an appointment and visa application fee*.

Cooper, C., & Hall, C. M. (2018). *Contemporary tourism: An international approach* (4th ed.). Oxford: Goodfellows Publishing.

Couldry, N. (2008). Mediatization or mediation? Alternative understandings of the emergent space of digital storytelling. *New Media & Society*, 10(3), 373–391.

Counted, V. (2016). Making sense of place attachment: Towards a holistic understanding of people-place relationships and experiences. *Environment, Space, Place*, 8(1), 7–32.

Counted, V. (2019). The role of spirituality in promoting sense of place among foreigners of African background in the Netherlands. *Ecopsychology*, 11(2), 101–109.

Counted, V., Moustafa, A., & Renzaho, A. (2019). Migration and sociodemographic factors associated with sense of place attitudes among migrants of African background in Northern and Western Netherlands. *Journal of Community Psychology*, 47(2), 238–253.

Counted, V., & Watts, F. (2017). Place attachment in the Bible: The role of attachment to sacred places in religious life. *Journal of Psychology and Theology*, 45(3), 218–232.

Counted, V., & Zock, H. (2019). Place spirituality: An attachment perspective. *Archive for the Psychology of Religion*, 41(1), 12–25.

Cox, R. S., & Perry, K. M. E. (2011). Like a fish out of water: Reconsidering disaster recovery and the role of place and social capital in community disaster resilience. *American Journal of Community Psychology*, 48(3–4), 395–411.

Craik, K. H., & Zube, E. H. (1976). The development of perceived environmental quality indices. In K. H. Craik (Ed.), *Perceiving environmental quality* (pp. 3–20). Boston, MA: Springer.

Cross, J. E. (2015). Processes of place attachment: An interactional framework. *Symbolic Interaction*, 38(4), 493–520.

Croy, W. G. (2010). Planning for film tourism: Active destination image management. *Tourism and Hospitality Planning & Development*, 7(1), 21–30.

Cruikshank, J. (1990). *Life lived like a story: Life stories of three Yukon Native elders*. Lincoln, NE: University of Nebraska Press.

Cuba, L., & Hummon, D. M. (1993). Constructing a sense of home: Place affiliation and-migration across the life cycle. *Sociological Forum*, 8, 547–572

Cuffee, Y., Ogedegbe, C., Williams, N. J., Ogedegbe, G., & Schoenthaler, A. (2014). Psychosocial risk factors for hypertension: An update of the literature. *Current Hypertension Reports*, 16(10), 483.

Cupples, J., & Glynn, K. (2009). Countercartographies: New (Zealand) cultural studies/geographies and the city. *New Zealand Geographer*, 65(1), 1–5.

Damjanov, K., & Crouch, D. (2018). Extra-planetary mobilities and the media prospects of virtual space tourism. *Mobilities*, 13(1), 1–13.

Danado, J., Dias, A. E., Romão, T., Correia, N., Trabuco, A., Santos, C., … Serpa, J. (2003). *Mobile augmented reality for environmental management (MARE)*. Paper presented at *24th Annual Conference of the European Association for Computer Graphics, Eurographics*, Granada, Spain, September 1–5.

Davey-Rothwell, M. A., German, D., & Latkin, C. A. (2008). Residential transience and depression: Does the relationship exist for men and women? *Journal of Urban Health*, 85(5), 707–716.

D'Andrea, A. (2006). Neo-nomadism: A theory of post-identitarian mobility in the global age. *Mobilities*, 1(1), 95–119.

de Bernardi, C. (2019). Authenticity as a compromise: A critical discourse analysis of Sámi tourism websites. *Journal of Heritage Tourism*, 14(3), 249–262.

Deffner, A., & Metaxas, T. (2010). Place marketing, local identity and branding cultural images in Southern Europe: Nea Ionia, Greece and Pafos, Cyprus. In G. Ashworth & M. Kavaratzis (Eds.), *Towards effective place brand management, branding European cities and regions* (pp. 49–68). Cheltenham, UK: Edward Elgar.

De Freitas, C. R., & Schmekal, A. (2006). Studies of condensation/evaporation processes in the Glowworm Cave, New Zealand. *International Journal of Speleology*, 35(2), 3.

Delaney, C. (1990). The Hajj: Sacred and secular. *American Ethnologist*, 17(3), 513–530.

Della Dora, V. (2012). Setting and blurring boundaries: Pilgrims, tourists, and landscape in Mount Athos and Meteora. *Annals of Tourism Research*, 39(2), 951–974.

Demiroglu, O. C., & Hall, C. M. (2020). Geobibliography and bibliometric networks of polar tourism and climate change research. *Atmosphere*, 11(5), 498.

Derr, V. (2002). Children's sense of place in northern New Mexico. *Journal of Environmental Psychology*, 22(1–2), 125–137.

Devine, K. (2017). *Sense of place: The phenomenology of virtual heritage place*. In *21st International Conference Information Visualisation (IV)*, London, 11–14 July, pp. 332–335, doi:10.1109/iV.2017.22

Devine-Wright, P. (2013). Think global, act local? The relevance of place attachments and place identities in a climate changed world. *Global Environmental Change*, 23(1), 61–69.

Devine-Wright, P., Price, J., & Leviston, Z. (2015). My country or my planet? Exploring the influence of multiple place attachments and ideological beliefs upon climate change attitudes and opinions. *Global Environmental Change*, 30, 68–79.

DeYoung, S. E., & Peters, M. (2016). My community, my preparedness: The role of sense of place, community, and confidence in government in disaster readiness. *International Journal of Mass Emergencies & Disasters*, 34(2), 250–282.

Dholakia, R. R., & Zhao, M. (2009). Retail web site interactivity. *International Journal of Retail & Distribution Management*, 37(10), 821–838.

Di Giovine, M. A., & Choe, J. (2019). Geographies of religion and spirituality: Pilgrimage beyond the 'officially' sacred. *Tourism Geographies*, 21(3), 361–383.

Di Masso, A., Dixon, J., & Durrheim, K. (2014). Place attachment as discursive practice. In L. C. Manzo & P. Devine-Wright (Eds.), *Place attachment: Advances in theory, methods and applications* (pp. 75–86). New York, NY: Routledge.

Diemer, J., Alpers, G. W., Peperkorn, H. M., Shiban, Y., & Mühlberger, A. (2015). The impact of perception and presence on emotional reactions: A review of research in virtual reality. *Frontiers in Psychology*, 6, 26. https://doi.org/10.3389/fpsyg.2015.00026

Dillimono, H. D., & Dickinson, J. E. (2015). Travel, tourism, climate change, and behavioral change: Travelers' perspectives from a developing country, Nigeria. *Journal of Sustainable Tourism*, 23(3), 437–454.

Dixon, J., & Durrheim, K. (2000). Displacing place-identity: A discursive approach to locating self and other. *British Journal of Social Psychology*, 39(1), 27–44.

Domínguez-Quintero, A. M., González-Rodríguez, M. R., & Paddison, B. (2020). The mediating role of experience quality on authenticity and satisfaction in the context of cultural-heritage tourism. *Current Issues in Tourism*, 23(2), 248–260.

Dominy, M. D. (2001). *Calling the station home: Place and identity in New Zealand's high country*. Lanham, MD: Rowman & Littlefield.

Doss, E. (1999). *Elvis culture: Fans, faith, and image*. Kansas City, KA: University Press of Kansas.

Dresner, M., Handelman, C., Braun, S., & Rollwagen-Bollens, G. (2015). Environmental identity, pro-environmental behaviors, and civic engagement of volunteer stewards in Portland area parks. *Environmental Education Research*, 21(7), 991–1010.

Du, H. (2017). Place attachment and belonging among educated young migrants and returnees: The case of Chaohu, China. *Population, Space and Place*, 23(1), e1967.

Dueck, A., & Byron, K. (2011). Community, spiritual traditions, and disasters in collective societies. *Journal of Psychology and Theology*, 39(3), 244–254.

Dueholm, J., & Smed, K. M. (2014). Heritage authenticities–a case study of authenticity perceptions at a Danish heritage site. *Journal of Heritage Tourism*, 9(4), 285–298.

Dunlop, F. (2006). The strange tale of General Tso's chicken. In R. Hosking (Ed.), *Authenticity in the kitchen* (pp. 165–177). Devon, UK: Prospect Books.

Dupre, K. (2019). Trends and gaps in place-making in the context of urban development and tourism: 25 years of literature review. *Journal of Place Management and Development*, 12(1), 102–120

Durán-Sánchez, A., Álvarez-García, J., Río-Rama, D., De la Cruz, M., & Oliveira, C. (2018). Religious tourism and pilgrimage: Bibliometric overview. *Religions*, 9(9), 249. https://doi.org/10.3390/rel9090249

Dwyer, L., Chen, N. C., & Lee, J. (2019). The role of place attachment in tourism research. *Journal of Travel & Tourism Marketing*, 36(5), 645–652.

Edgerton, S. H. (1996). *Translating the curriculum: Multiculturalism into cultural studies*. New York, NY: Routledge.

Ehn, B., & Löfgren, O. (2010). *The secret world of doing nothing*. Berkeley, CA: University of California Press.

Elliot, S., Papadopoulos, N., & Kim, S. S. (2011). An integrative model of place image: Exploring relationships between destination, product, and country images. *Journal of Travel Research*, 50(5), 520–534.

Ellis, N. R., & Albrecht, G. A. (2017). Climate change threats to family farmers' sense of place and mental wellbeing: A case study from the Western Australian Wheatbelt. *Social Science & Medicine*, 175, 161–168.

Elman, B. A. (1989). Imperial politics and Confucian societies in late imperial China: The Hanlin and Donglin academies. *Modern China*, 15(4), 379–418.

Engeset, A. B., & Heggem, R. (2015). Strategies in Norwegian farm tourism: Product development, challenges, and solutions. *Scandinavian Journal of Hospitality and Tourism*, 15(1–2), 122–137.

Engeset, M. G., & Elvekrok, I. (2015). Authentic concepts: Effects on tourist satisfaction. *Journal of Travel Research*, 54(4), 456–466.

Ettinger, A., Grabner-Kräuter, S., Okazaki, S., & Terlutter, R. (2020). The desirability of CSR communication versus greenhushing in the hospitality industry: The customers' perspective. *Journal of Travel Research*. https://doi.org/10.1177/0047287520930087.

Evans, K. (2017). Rock stars. *New Zealand Geographic*, 145. Retrieved from https://www.nzgeo.com/stories/rock-stars/

Falconer, L. (2017). Experiencing sense of place in virtual and physical Avebury. *Personal and Ubiquitous Computing*, 21(6), 977–988.

Falk, M., & Vieru, M. (2019). International tourism demand to Finnish Lapland in the early winter season. *Current Issues in Tourism*, 22(11), 1312–1326.

Farrar, M. E. (2011). Amnesia, nostalgia, and the politics of place memory. *Political Research Quarterly*, 64(4), 723–735.

Fein, J. (2006), Tapping the power of your ancestors. *Spirituality and Health*, July, 36–39.

Ferguson, M. R. (2011). Finding a home abroad with "Eveline": Using narrative inquiry to establish a sense of place for a Western teacher in a foreign and multicultural context. *Journal of Studies in International Education*, 15(1), 25–40.

Fink, C., & Miguélez, E. (Eds.). (2017). *The international mobility of talent and innovation: New evidence and policy implications*. Cambridge, UK: Cambridge University Press.

Finsterwalder, J., & Hall, C. M. (2016). Disasters, urban regeneration and the temporality of servicescapes. In C. M. Hall, S. Malinen, R. Vosslamber, & R. Wordsworth (Eds.), *Business and post-disaster management: Business, organisational and consumer resilience and the Christchurch earthquakes* (pp. 230–248). Oxford, UK: Routledge.

Fiore, A. M., Kim, J., & Lee, H. H. (2005). Effect of image interactivity technology on consumer responses toward the online retailer. *Journal of Interactive Marketing*, 19(3), 38–53.

Fisher, W. R. (1984). Narration as a human communication paradigm: The case of public moral argument. *Communications Monographs*, 51(1), 1–22.

Forristal, L. J., Lehto, X. Y., & Lee, G. (2014). The contribution of native species to sense of place. *Current Issues in Tourism*, 17(5), 414–433.

Frost, W., Frost, J., Strickland, P., & Maguire, J. S. (2020). Seeking a competitive advantage in wine tourism: Heritage and storytelling at the cellar-door. *International Journal of Hospitality Management*, 87, 102460.

Fuhrer, U., Kaiser, F. G., & Hartig, T. (1993). Place attachment and mobility during leisure time. *Journal of Environmental Psychology*, 13(4), 309–321.

Gabriel, Y. (2000). *Storytelling in organizations: Facts, fictions, and fantasies*. Oxford, UK: Oxford University Press.

Galegos, D. (2011). Social approaches to understanding food, eating and nutrition. In M. L. Wahlqvist (Ed.), *Food and nutrition: Food and health systems in Australia and New Zealand* (pp. 21–35). Sydney: Allen & Unwin.

Gangjee, D. S. (Ed.). (2016). *Research handbook on intellectual property and geographical indications*. Cheltenham, UK: Edward Elgar.

Garfinkel, M. (2019). Place matters: Rooting conservation in community. *The Journal of American Folklore*, 132(526), 412–430.

Ghaderi, Z., Hall, C. M., Scott, N., & Béal, L. (2020). Islamic beliefs and host-guest relationships in Iran. *International Journal of Hospitality Management*, 90. https://doi.org/10.1016/j.ijhm.2020.102603

Gardiner, S., & Kwek, A. (2017). Chinese participation in adventure tourism: A study of generation Y international students' perceptions. *Journal of Travel Research*, 56(4), 496–506.

George, L. (2010). The interweaving of people, time, and place–Whakapapa as context and method. *Pacific Studies*, 33(2), 241–258.

Ghani, I., Rafi, A., & Woods, P. (2019). The effect of immersion towards place presence in virtual heritage environments. *Personal and Ubiquitous Computing*. https://doi.org/10.1007/s00779-019-01352-8

Gibson, J., Rohorua, H., McKenzie, D., & Stillman, S. (2010). Information flows and migration: Recent survey evidence from the South Pacific. *Asian and Pacific Migration Journal*, 19(3), 401–420.

Giroux, H. A., & McLaren, P. (1994). *Between borders: Pedagogy and the politics of cultural studies*. New York, NY: Routledge.

Goel, L., Johnson, N. A., Junglas, I., & Ives, B. (2011). From space to place: Predicting users' intentions to return to virtual worlds. *MIS Quarterly*, 35(3), 749–771.

Goh, C., & Law, R. (2002). Modeling and forecasting tourism demand for arrivals with stochastic nonstationary seasonality and intervention. *Tourism Management*, 23(5), 499–510.

Golden, C. (2014). Spiritual roots of the land: Hierarchy and relationships of the religious cosmologies of humans and their environment in the Maroantsetra region of Madagascar. *Worldviews: Global Religions, Culture, and Ecology*, 18(3), 255–268.

González-Rodríguez, M. R., Díaz-Fernández, M. C., & Pino-Mejías, M. Á. (2020). The impact of virtual reality technology on tourists' experience: A textual data analysis. *Soft Computing*, 24, 13879–13892.

Goodsell, T. L., & Seiter, L. (2011). Scrapbooking: Family capital and the construction of family discourse. *Journal of Contemporary Ethnography*, 40(3), 318–341.

Goudy, W. J. (1990). Community attachment in a rural region. *Rural Sociology*, 55(2), 178–198.

Greene, E. (1996). *Planet of the Apes as American myth: Race and politics in the films and television series*. Jefferson, NC: McFarland.

Grenier, A. A. (2007). The diversity of polar tourism. Some challenges facing the industry in Rovaniemi, Finland. *Polar Geography*, 30(1–2), 55–72.

Grenni, S., Horlings, L. G., & Soini, K. (2020). Linking spatial planning and place branding strategies through cultural narratives in places. *European Planning Studies*, 28(7), 1355–1374.

Grigg, D. (1995). The geography of food consumption: A review. *Progress in Human Geography*, 19(3), 338–354.

Groshong, L., Stanis, S. W., Morgan, M., & Li, C. J. (2020). Place attachment, climate friendly behavior, and support for climate friendly management action among state park visitors. *Environmental Management*, 65(1), 98–110.

Gu, H., & Ryan, C. (2008). Place attachment, identity and community impacts of tourism—the case of a Beijing hutong. *Tourism Management*, 29(4), 637–647.

Guo, Y., Zhang, J., Zhang, Y., & Zheng, C. (2018). Catalyst or barrier? The influence of place attachment on perceived community resilience in tourism destinations. *Sustainability*, 10(7), 2347. https://doi.org/10.3390/su10072347

Gurney, G. G., Blythe, J., Adams, H., Adger, W. N., Curnock, M., Faulkner, L., James, T., & Marshall, N. A. (2017). Redefining community based on place attachment in a connected world. *Proceedings of the National Academy of Sciences*, 114(38), 10077–10082.

Gutierrez, M., Vexo, F., & Thalmann, D. (2008). *Stepping into virtual reality*. London: Springer Science & Business Media.

Guttentag, D. A. (2010). Virtual reality: Applications and implications for tourism. *Tourism Management*, 31(5), 637–651.

Gössling, S., McCabe, S., & Chen, N. C. (2020). A socio-psychological conceptualisation of overtourism. *Annals of Tourism Research*, 84. https://doi.org/10.1016/j.annals.2020.102976

Hagerty, B. M., Lynch-Sauer, J., Patusky, K. L., Bouwsema, M., & Collier, P. (1992). Sense of belonging: A vital mental health concept. *Archives of Psychiatric Nursing*, 6(3), 172–177.

Hall, C. M. (1996) Wine tourism in New Zealand. In Centre for Tourism (Ed.), *Proceedings of Tourism Down Under II: A Research Conference* (pp. 109–119). Dunedin: University of Otago.

Hall, C. M. (1997). Geography, marketing and the selling of places. *Journal of Travel & Tourism Marketing*, 6(3–4), 61–84.

Hall, C. M. (2002). ANZAC Day and secular pilgrimage. *Tourism Recreation Research*, 27(2), 87–91.

Hall, C. M. (2005). *Tourism: Rethinking the social science of mobility*. Harlow: Person.

Hall, C. M. (2006a). Travel and journeying on the Sea of Faith: Perspectives from Religious Humanism. In D. Timothy and D. Olsen (Eds.), *Tourism, religion and spiritual journeys* (pp. 64–77). New York: Routledge.

Hall, C. M. (2006b). New Zealand tourism entrepreneur attitudes and behaviours with respect to climate change adaptation and mitigation. *International Journal of Innovation and Sustainable Development*, 1(3), 229–237.

Hall, C. M. (2007). Response to Yeoman et al: The fakery of 'The authentic tourist'. *Tourism Management*, 28(4), 1139–1140.

Hall, C. M. (2008a). Santa Claus, place branding and competition. *Fennia-International Journal of Geography*, 186(1), 59–67.

Hall, C. M. (2008b). Servicescapes, designscapes, branding, and the creation of place-identity: South of Litchfield, Christchurch. *Journal of Travel & Tourism Marketing*, 25(3–4), 233–250.

Hall, C. M. (2009). Heritage tourism in the Pacific: Modernity, myth, and identity. In D. Timothy (Ed.), *Cultural heritage and tourism in the developing world* (pp. 123–140). Abingdon, UK: Routledge.

Hall, C. M. (2011). Consumerism, tourism and voluntary simplicity: We all have to consume, but do we really have to travel so much to be happy? *Tourism Recreation Research*, 36(3), 298–303.

Hall, C. M. (2013). Glow-worm tourism in Australia and New Zealand: Commodifying and conserving charismatic micro-fauna. In R. H. Lemelin (Ed.), *The management of insects in recreation and tourism* (pp. 217–232). Cambridge: Cambridge University Press.

Hall, C. M. (2014a). Will climate change kill Santa Claus? Climate change and high-latitude Christmas place branding. *Scandinavian Journal of Hospitality and Tourism*, 14(1), 23–40.

Hall, C. M. (2014b). Second home tourism: An international review. *Tourism Review International*, 18(3), 115–135.

Hall, C. M. (2015). Tourism planning and human security. In R. Nunkoo & S. Smith (Eds.), *Trust, tourism development and planning* (pp. 86–110). Abingdon, UK: Routledge.

Hall, C. M., & Frost, W. (2009). National parks and the 'worthless lands hypothesis' revisited. In W. Frost & C. M. Hall (Eds.), *Tourism and national parks: International perspectives on development, histories and change* (pp. 45–62). London: Routledge.

Hall, C. M., & Macionis, N. (1998). Wine tourism in Australia and New Zealand. In R. W. Butler, C. M. Hall and J. Jenkins (Eds.), *Tourism and recreation in rural areas* (pp. 197–224). Chichester, UK: John Wiley & Sons.

Hall, C. M., Malinen, S., Vosslamber, R., & Wordsworth, R. (Eds.). (2016). *Business and post-disaster management: Business, organisational and consumer resilience and the Christchurch earthquakes*. Abingdon, UK: Routledge.

Hall, C. M., Mitchell, I., & Keelan, N. (1992). Maori culture and heritage tourism in New Zealand. *Journal of Cultural Geography*, 12(2), 115–128.

Hall, C. M., Mitchell, I., & Keelan, N. (1993). The implications of Maori perspectives for the management and promotion of heritage tourism in New Zealand. *GeoJournal*, 29(3), 315–322.

Hall, C. M., & Mitchell, R. (2002). The tourist terror of New Zealand wine: The importance of region in the wine tourism experience. In A. Montanari (Ed.), *Food and environment: Geographies of taste* (pp. 69–91). Rome, Italy: Società Geografica Italiana.

Hall, C. M., & Mitchell, R. (2008). *Wine marketing*. Oxford, UK: Elsevier.

Hall, C. M., Mitchell, R., & Sharples, L. (2004a). Consuming places: The role of food, wine and tourism in regional development. In C. M. Hall, L. Sharples, R. Mitchell, N.

Macionis, & B. Cambourne (Eds.), *Food tourism around the world* (pp. 25–59). Oxford, UK: Routledge.

Hall, C. M., & Müller, D. (2018). Governing and planning for second homes. In C. M. Hall & D. Müller (Eds.), *The Routledge handbook of second home tourism and mobilities* (pp. 17–26). Abingdon, UK: Routledge.

Hall, C. M., & Page, S. J. (2014). *The geography of tourism and recreation* (4th ed.). Abingdon, UK: Routledge.

Hall, C. M., & Prayag, G. (Eds.). (2020). *The Routledge handbook of halal hospitality and Islamic tourism*. Abingdon, UK: Routledge.

Hall, C. M., & Saarinen, J. (2020). 20 Years of Nordic climate change crisis and tourism research: A review and future research agenda. *Scandinavian Journal of Hospitality and Tourism*. https://doi.org/10.1080/15022250.2020.1823248.

Hall, C. M., & Seyfi, S. (Eds.). (2020). *Cultural and heritage tourism in the Middle East and North Africa: Complexities, management and practices*. Abingdon, UK: Routledge.

Hall, C. M., Sharples, L., Mitchell, R., Macionis, N., & Cambourne, B. (Eds.) (2004b). *Food tourism around the world*. Oxford, UK: Elsevier.

Hall, C. M., & Tucker, H. (Eds.). (2004). *Tourism and postcolonialism contested discourses, identities and representation*. Oxford, UK: Routledge.

Hall, C. M., & Williams, A. M. (Eds.). (2002). *Tourism and migration: New relationships between production and consumption*. Dordtrecht: Kluwer.

Hall, C. M., & Williams, A. M. (2020). *Tourism and innovation* (2nd ed.). Abingdon: Routledge.

Halliday, J. (1998). *Wine atlas of Australia & New Zealand*. Sydney, NSW: Harper Collins.

Halpenny, E. A. (2010). Pro-environmental behaviours and park visitors: The effect of place attachment. *Journal of Environmental Psychology*, 30(4), 409–421.

Hart, R. (1979). *Children's experience of place*. New York, NY: Irvington.

Hassan, A. (2016). Electronic word-of-mouth: An industry tailored application for tourism promotion. In S. Rathore & A. Panwar (Eds.), *Capturing, analyzing, and managing word-of-mouth in the digital marketplace* (pp. 61–75). Hershey, PA: IGI Global.

Hay, R. (1998a). Sense of place in developmental context. *Journal of Environmental Psychology*, 18(1), 5–29.

Hay, R. (1998b). A rooted sense of place in cross-cultural perspective. *Canadian Geographer/Le Géographe Canadien*, 42(3), 245–266.

Herbert, D. (2001). Literary places, tourism and the heritage experience. *Annals of Tourism Research*, 28(2), 312–333.

Hernández, B., Martín, A. M., Ruiz, C., & del Carmen Hidalgo, M. (2010). The role of place identity and place attachment in breaking environmental protection laws. *Journal of Environmental Psychology*, 30(3), 281–288.

Hibbert, J. F., Dickinson, J. E., & Curtin, S. (2013a). Understanding the influence of interpersonal relationships on identity and tourism travel. *Anatolia*, 24(1), 30–39.

Hibbert, J. F., Dickinson, J. E., Gössling, S., & Curtin, S. (2013b). Identity and tourism mobility: An exploration of the attitude–behaviour gap. *Journal of Sustainable Tourism*, 21(7), 999–1016.

Higginbotham, G. (2012). Seeking roots and tracing lineages: Constructing a framework of reference for roots and genealogical tourism. *Journal of Heritage Tourism*, 7(3), 189–203.

Hiltunen, M. J., Pitkänen, K., Vepsäläinen, M., & Hall, C. M. (2013). Secondhome tourism in Finland-Current trends and eco-social impacts. In Z. Roca (Ed.), *Second home tourism in Europe: Lifestyle issues and policy responses* (pp. 165–200). Aldershot, UK: Ashgate.

Hoersting, R. C., & Jenkins, S. R. (2011). No place to call home: Cultural homelessness, self-esteem and cross-cultural identities. *International Journal of Intercultural Relations*, 35(1), 17–30.

Hoey, B. A. (2010). Place for personhood: Individual and local character in lifestyle migration. *City & Society*, 22(2), 237–261.

Holtorf, C. (2013). On pastness: A reconsideration of materiality in archaeological object authenticity. *Anthropological Quarterly*, 86(2), 427–443.

Horton, G., Hanna, L., & Kelly, B. (2010). Drought, drying and climate change: Emerging health issues for ageing Australians in rural areas. *Australasian Journal on Ageing*, 29(1), 2–7.

Hosany, S., Prayag, G., Van Der Veen, R., Huang, S., & Deesilatham, S. (2017). Mediating effects of place attachment and satisfaction on the relationship between tourists' emotions and intention to recommend. *Journal of Travel Research*, 56(8), 1079–1093.

Hsu, J. L., Lu, H. C. H., & Chen, C. M. L. (2014). The effect of family beef taboos on beef consumption on young Taiwanese adults. *Food Quality and Preference*, 34, 45–49.

Hsu, S. Y., Dehuang, N., & Woodside, A. G. (2009). Storytelling research of consumers' self-reports of urban tourism experiences in China. *Journal of Business Research*, 62(12), 1223–1254.

Huang, W. J., & Chen, C. C. (2020). Influence of transnational leisure on diaspora tourism among contemporary migrants. *Journal of Travel Research*. https://doi.org/10.1177/0047287520922315

Huang, W. J., King, B., & Suntikul, W. (2017). VFR tourism and the tourist gaze: Overseas migrant perceptions of home. *International Journal of Tourism Research*, 19(4), 421–434.

Huang, W. J., Ramshaw, G., & Norman, W. C. (2016). Homecoming or tourism? Diaspora tourism experience of second-generation immigrants. *Tourism Geographies*, 18(1), 59–79.

Huang, Y. (2011). Identity negotiation in relation to context of communication. *Theory and Practice in Language Studies*, 1(3), 219–225.

Huang, X., Zhao, B., Liu, Y., & Xue, D. (2020). Belonging to a place: An analysis of the perceptions of rural-to-urban migrants in China. *Geographical Review*, 110(3), 406–424.

Hudson, M. L., Ahuriri-Driscoll, A. L., Lea, M. G., & Lea, R. A. (2007). Whakapapa–a foundation for genetic research? *Journal of Bioethical Inquiry*, 4(1), 43–49.

Hudson, S., & Ritchie, J. B. (2006). Promoting destinations via film tourism: An empirical identification of supporting marketing initiatives. *Journal of Travel Research*, 44(4), 387–396.

Huff, L. A. (2006). Sacred sustenance: Maize, storytelling, and a Maya sense of place. *Journal of Latin American Geography*, 5(1), 79–96.

Hughes, N. (2018). 'Tourists go home': Anti-tourism industry protest in Barcelona. *Social Movement Studies*, 17(4), 471–477.

Hultman, J., & Hall, C. M. (2012). Tourism place-making: Governance of locality in Sweden. *Annals of Tourism Research*, 39(2), 547–570.

Hummon, D. M. (1992). Community attachment. In S. M. Low & I. Altman (Eds.), *Place attachment* (pp. 253–278). Boston, MA: Springer.

Husa, A., & Morse, C. E. (2020). Rurality as a key factor for place attachment in the great plains. *Geographical Review*. https://doi.org/10.1080/00167428.2020.1786384

Hussain, Y. (2019). 'I know my roots are Indian but my thinking is Kiwi': Hybridisation, identity and 'Indians' in New Zealand. *South Asian Diaspora*, 11(1), 1–15.

Hyun, M. Y., & O'Keefe, R. M. (2012). Virtual destination image: Testing a telepresence model. *Journal of Business Research*, 65(1), 29–35.

Iaquinto, B. L. (2020). Understanding the place-making practices of backpackers. *Tourist Studies*. https://doi.org/10.1177/1468797620927308

IJsselsteijn, W. A., De Ridder, H., Freeman, J., & Avons, S. E. (2000). *Presence: Concept, determinants, and measurement. Proceedings of the SPIE, Human Vision and Electronic Imaging V*, 2 June 2000 (Vol. 3959, pp. 520–529). San Jose, CA: International Society for Optics and Photonics.

Inalhan, G., & Finch, E. (2004). Place attachment and sense of belonging in the offices. *Facilities*, 22(5/6), 120–128.

Iyer, P. (1999). Foreword: A journey into candlelight. In B. Bouldrey (ed.) *Travelling souls: Contemporary pilgrimage stories* (pp. vii–xii), San Francisco, CA: Whereabouts Press.

Jack, G. (2010). Place matters: The significance of place attachments for children's well-being. *British Journal of Social Work*, 40(3), 755–771.

Jackson, J. B. (1994). *A sense of place, a sense of time*. New Haven, CT: Yale University Press.

Jacobson, D. (1996). *Rights across borders: Immigration and the decline of citizenship*. Baltimore, MD: Johns Hopkins University Press.

Janta, H., & Christou, A. (2019). Hosting as social practice: Gendered insights into contemporary tourism mobilities. *Annals of Tourism Research*, 74, 167–176.

Jeffers, J. M. (2019). Saving Cork City? Place attachment and conflicting framings of flood hazards. *Geoforum*, 103, 26–35.

Jeffreys, B. (2020). UK universities see boom in Chinese students. *BBC News*, 21 January. Retrieved from https://www.bbc.com/news/education-51149445

Jenkins, H. (2004). The cultural logic of media convergence. *International Journal of Cultural Studies*, 7(1), 33–43.

Jenkins, O. (2003). Photography and travel brochures: The circle of representation. *Tourism Geographies*, 5(3), 305–328.

Jewell, B., & McKinnon, S. (2008). Movie tourism—A new form of cultural landscape? *Journal of Travel & Tourism Marketing*, 24(2–3), 153–162.

Jindra, M. (1994). Star Trek fandom as a religious phenomenon. *Sociology of Religion*, 55(1), 27–51.

Johnstone, B. (1990). *Stories, community, and place: Narratives from Middle America*. Bloomington, IN: Indiana University Press.

Joppke, C. (2007). Transformation of citizenship: Status, rights, identity. *Citizenship Studies*, 11(1), 37–48.

Jorgensen, B. S., & Stedman, R. C. (2001). Sense of place as an attitude: Lakeshore owners attitudes toward their properties. *Journal of Environmental Psychology*, 21(3), 233–248.

Judd, D. R. (1995). Promoting tourism in US cities. *Tourism Management*, 16(3), 175–187.

Kajan, E. (2014). Community perceptions to place attachment and tourism development in Finnish Lapland. *Tourism Geographies*, 16(3), 490–511.

Kalandides, A. (2011). The problem with spatial identity: Revisiting the "sense of place". *Journal of Place Management and Development*, 4(1), 28–39.

Kaltenborn, B. P. (1998). Effects of sense of place on responses to environmental impacts: A study among residents in Svalbard in the Norwegian high Arctic. *Applied Geography*, 18(2), 169–189.

Karahalios, K., & Donath, J. (2004). Telemurals: Linking remote spaces with social catalysts. In *Proceedings of the SIGCHI conference on Human factors in computing systems*, April (pp. 615–622). ACM Digital Library, https://doi.org/10.1145/985692.985770

Kashiwagi, S., Nagai, H., & Furutani, T. (2018). VFR travel generated by international students: The case of Japanese students in Australia. *Tourism*, 66(1), 89–103.

Kavaratzis, M., & Ashworth, G. (2008). Place marketing: How did we get here and where are we going? *Journal of Place Management and Development*, 1, 150–165.

Kenny, G. J., Harman, J. J., Flux, T. L., Warrick, R. A., & Ye, W. (2001). The impact of climate change on regional resources: A case study for Canterbury and Waikato regions. In R. A. Warrick, G. J. Kenny, & J. J. Harman (Eds.), *The effects of climate change and variation in New Zealand: An assessment using the CLIMPACTS system* (pp. 85–98). Hamilton, New Zealand: International Global Change Institute (IGCI), University of Waikato.

Khanian, M., Serpoush, B., & Gheitarani, N. (2019). Balance between place attachment and migration based on subjective adaptive capacity in response to climate change: The case of Famenin County in Western Iran. *Climate and Development*, 11(1), 69–82.

Kim, B., Kim, S., & King, B. (2020a). Religious tourism studies: Evolution, progress, and future prospects. *Tourism Recreation Research*, 45(2), 185–203.

Kim, M. J., & Hall, C. M. (2019). A hedonic motivation model in virtual reality tourism: Comparing visitors and non-visitors. *International Journal of Information Management*, 46, 236–249.

Kim, M. J., & Hall, C. M. (2020). What drives visitor economy crowdfunding? The effect of digital storytelling on unified theory of acceptance and use of technology. *Tourism Management Perspectives*, 34, 100638.

Kim, M. J., Lee, C. K., & Jung, T. (2020b). Exploring consumer behavior in virtual reality tourism using an extended stimulus-organism-response model. *Journal of Travel Research*, 59(1), 69–89.

King, B., & Gardiner, S. (2015). Chinese international students. An avant-garde of independent travellers? *International Journal of Tourism Research*, 17(2), 130–139.

Kirschenbaum, A. (2002). Disaster preparedness: A conceptual and empirical reevaluation. *International Journal of Mass Emergencies and Disasters*, 20(1), 5–28.

Kirillova, K. (2019). Existentialism and tourism: New research avenues. *International Journal of Tourism Cities*, 5(3), 429–442.

Kirillova, K., & Lehto, X. (2015). An existential conceptualization of the vacation cycle. *Annals of Tourism Research*, 55, 110–123.

Klein, L. R. (2003). Creating virtual product experiences: The role of telepresence. *Journal of Interactive Marketing*, 17(1), 41–55.

Ko Tāne. (n.d.). *Homepage*. Retrieved from http://www.kotane.co.nz/

Kolar, T., & Zabkar, V. (2010). A consumer-based model of authenticity: An oxymoron or the foundation of cultural heritage marketing? *Tourism Management*, 31(5), 652–664.

Kotler, P., & Gertner, D. (2002). Country as brand, product, and beyond: A place marketing and brand management perspective. *Journal of Brand Management*, 9(4), 249–261.

Kruse, R. J. (2003). Imagining Strawberry Fields as a place of pilgrimage. *Area*, 53(2), 154–162.

Kruse, R. J. (2005). The Beatles as place makers: Narrated landscapes in Liverpool, England. *Journal of Cultural Geography*, 22(2), 87–114.

Kukkakorpi, M., & Pantti, M. (2020). A sense of place: VR journalism and emotional engagement. *Journalism Practice*. https://doi.org/10.1080/17512786.2020.1799237

Kyle, G., Graefe, A., & Manning, R. (2005). Testing the dimensionality of place attachment in recreational settings. *Environment and Behavior*, 37(2), 153–177.

Kymlicka, W. (2001). *Politics in the vernacular: Nationalism, multiculturalism, and citizenship*. Oxford, UK: Oxford University Press.

Labrador, R. N. (2002). Performing identity: The public presentation of culture and ethnicity among Filipinos in Hawai'i. *Cultural Values*, 6(3), 287–307.

Lachapelle, P. (2008). A sense of ownership in community development: Understanding the potential for participation in community planning efforts. *Community Development*, 39(2), 52–59.

Laing, A., Newholm, T., & Hogg, G. (2009). Contradictory spaces: Negotiating virtual spaces of consumption. *Journal of Service Management*, 20(2), 227–245.

Lalli, M. (1992). Urban-related identity: Theory, measurement, and empirical findings. *Journal of Environmental Psychology*, 12(4), 285–303.

Larsen, J. (2006). Picturing Bornholm: Producing and consuming a tourist place through picturing practices. *Scandinavian Journal of Hospitality and Tourism*, 6(2), 75–94.

Lawson-Te Aho, K., & Liu, J. H. (2010). Indigenous suicide and colonization: The legacy of violence and the necessity of self-determination. *International Journal of Conflict and Violence*, 4(1), 124–133.

Le, M. T., Kim, H. K., & Kwon, Y. M. (2005). 3D modeling and adaptation for virtual heritage system. In G. Bebis, R. Boyle, D. Koracin, & B. Parvin (Eds.), *Advances in visual computing. ISVC 2005. Lecture notes in computer science, 3804*, (pp. 579–586). Berlin, Heidelberg: Springer.

Lee, J. (1978). Migration and expansion in Chinese history. In W. H. McNeill & R. S. Adams (Eds.), *Human migration: Patterns and policies* (pp. 20–47). Bloomington, IN: Indiana University Press.

Lee, J., Kyle, G., & Scott, D. (2012). The mediating effect of place attachment on the relationship between festival satisfaction and loyalty to the festival hosting destination. *Journal of Travel Research*, 51(6), 754–767.

Lee, K., & Zhao, J. (2003). Japanese travelers' service preferences in US hotels. *Journal of Travel & Tourism Marketing*, 14(2), 67–85.

Lee, K., Madanoglu, M., & Ko, J-Y. (2016). Exploring key service quality dimensions at a winery from an emerging market's perspective. *British Food Journal*, 118(12), 2981–2996.

Lee, K. M. (2004). Why presence occurs: Evolutionary psychology, media equation, and presence. *Presence: Teleoperators & Virtual Environments*, 13(4), 494–505.

Lee, M., Lee, S. A., Jeong, M., & Oh, H. (2020). Quality of virtual reality and its impacts on behavioral intention. *International Journal of Hospitality Management*, 90, 102595. https://doi.org/10.1016/j.ijhm.2020.102595

Lee, M. R., & Blanchard, T. C. (2012). Community attachment and negative affective states in the context of the BP Deepwater Horizon disaster. *American Behavioral Scientist*, 56(1), 24–47.

Lee, Y. S., & Shin, W. J. (2015). Marketing tradition-bound products through storytelling: A case study of a Japanese sake brewery. *Service Business*, 9(2), 281–295.

Leeman, S. (2006). 'Da Vinci Code' tourists flock to Scottish chapel. *NBC News*. http://www.nbcnews.com/id/12940481/ns/travel-destination_travel/t/da-vinci-code-tourists-flock-scottish-chapel/#.X1GQc8gzZPY

Lengen, C., & Kistemann, T. (2012). Sense of place and place identity: Review of neuroscientific evidence. *Health & Place*, 18(5), 1162–1171.

Lew, A. A. (2017). Tourism planning and place making: Place-making or placemaking?, *Tourism Geographies*, 19(3), 448–466.

Lewis, J. L., & Sheppard, S. R. (2005). Ancient values, new challenges: Indigenous spiritual perceptions of landscapes and forest management. *Society and Natural Resources*, 18(10), 907–920.

Lichrou, M., O'Malley, L., & Patterson, M. (2008). Place-product or place narrative (s)? Perspectives in the marketing of tourism destinations. *Journal of Strategic Marketing*, 16(1), 27–39.

Lichrou, M., O'Malley, L., & Patterson, M. (2010). Narratives of a tourism destination: Local particularities and their implications for place marketing and branding. *Place Branding and Public Diplomacy*, 6(2), 134–144.

Lichrou, M., O'Malley, L., & Patterson, M. (2014). On the marketing implications of place narratives. *Journal of Marketing Management*, 30(9–10), 832–856.

Li, C. W. C. (2016). The China that was not to be: Nostalgia and the exilic imagination of Taiwan's mainland diaspora. *Diaspora Studies*, 9(1), 1–14.

Li, F., & Maher, M. L. (2000). Representing virtual places – A design model for metaphorical design. In M. J. Clayton & G. de Velasco (Eds.), *Eternity, infinity and virtuality in architecture* (pp. 103–111). Washington, DC: Acadia.

Li, T. E., & Chan, E. T. H. (2018). Connotations of ancestral home: An exploration of place attachment by multiple generations of Chinese diaspora. *Population, Space and Place*, 24(8), e2147. https://doi.org/10.1002/psp.2147

Li, T. E., McKercher, B., & Chan, E. T. H. (2020). Towards a conceptual framework for diaspora tourism. *Current Issues in Tourism*. https://doi.org/10.1080/13683500.2019.1634013

Liang, J. L., & Elliot, S. (2020). A systematic review of augmented reality tourism research: What is now and what is next? *Tourism and Hospitality Research*, https://doi.org/10.1177%2F1467358420941913

Lin, J., Guia Julve, J., Xu, H., & Cui, Q. (2020a). Food habits and tourist food consumption: An exploratory study on dining behaviours of Chinese outbound tourists in Spain. *Journal of Policy Research in Tourism, Leisure and Events*, 12(1), 82–99.

Lin, L. P. L., Huang, S. C. L., & Ho, Y. C. (2020b). Could virtual reality effectively market slow travel in a heritage destination? *Tourism Management*, 78. https://doi.org/10.1016/j.tourman.2019.104027

Liu, C. X. (2017). *The wandering earth*. Wuhan, China: Changjiang Literature and Art Publishing House.

Liu, S. (2015). Searching for a sense of place: Identity negotiation of Chinese immigrants. *International Journal of Intercultural Relations*, 46, 26–35.

Lombard, M., & Ditton, T. (1997). At the heart of it all: The concept of presence. *Journal of Computer-mediated Communication*, 3(2). https://doi.org/10.1111/j.1083-6101.1997.tb00072.x

Loureiro, S. M. C., Guerreiro, J., & Ali, F. (2020). 20 years of research on virtual reality and augmented reality in tourism context: A text-mining approach. *Tourism Management*, 77. https://doi.org/10.1016/j.tourman.2019.104028

Low, S. M., & Altman, I. (1992). *Place attachment*. Boston, MA: Springer.

Lowe, J., & DeVerteuil, G. (2020). Power, powerlessness and the politics of mobility: Reconsidering mental health geographies. *Social Science & Medicine*, 252, 112918.

Lowenthal, D. (1975). Past time, present place: Landscape and memory. *Geographical Review*, 65, 1–36.

Lowenthal, D. (1985). *The past is a foreign country*. Cambridge: Cambridge University Press.

Lundqvist, A., Liljander, V., Gummerus, J., & Van Riel, A. (2013). The impact of storytelling on the consumer brand experience: The case of a firm-originated story. *Journal of Brand Management*, 20(4), 283–297.

Lutts, R. H. (1985). Place, home, and story in environmental education. *The Journal of Environmental Education*, 17(1), 37–41.

Ma, Y., Ooi, C. S., & Hardy, A. (2018). Chinese travelling overseas and their anxieties. In E. C. L. Yang & C. Khoo-Lattimore (Eds.), *Asian cultures and contemporary tourism* (pp. 201–220). Singapore: Springer.

MacDonald, J. (1998). *The Arctic sky: Inuit astronomy, star lore, and legend.* Toronto, CA: Royal Ontario Museum and Nunavut Research Institute.

Mair, H. (2006). Global restructuring and local responses: Investigating rural tourism policy in two Canadian communities. *Current Issues in Tourism*, 9(1), 1–45.

Malatzky, C., Gillespie, J., Couch, D. L., & Cosgrave, C. (2020). Why place matters: A rurally-orientated analysis of COVID-19's differential impacts. *Social Sciences & Humanities Open*, 2(1), 100063.

Malpas, J. (2008). New media, cultural heritage and the sense of place: Mapping the conceptual ground. *International Journal of Heritage Studies*, 14(3), 197–209.

Mansilla, J. A., & Milano, C. (2019). Becoming centre: Tourism placemaking and space production in two neighborhoods in Barcelona. *Tourism Geographies*. https://doi.org/10.1080/14616688.2019.1571097.

Marschall, S. (2014). Tourism and remembrance: The journey into the self and its past. *Journal of Tourism and Cultural Change*, 12(4), 335–348.

Marschall, S. (2015). 'Homesick tourism': Memory, identity and (be) longing. *Current issues in Tourism*, 18(9), 876–892.

Marshall, J. (2001). Women and strangers: Issues of marginalization in seasonal tourism. *Tourism Geographies*, 3(2), 165–186.

Marshall, J., & Foster, N. (2002). "Between belonging": Habitus and the migration experience. *Canadian Geographer/Le Géographe Canadien*, 46(1), 63–83.

Marshall, N., Adger, W. N., Benham, C., Brown, K., Curnock, M. I., Gurney, G. G., Marshall, P., Pert, P. L., & Thiault, L. (2019). Reef grief: Investigating the relationship between place meanings and place change on the Great Barrier Reef, Australia. *Sustainability Science*, 14(3), 579–587.

Martin-Jones, D. (2014). Film tourism as heritage tourism: Scotland, diaspora and *The Da Vinci Code* (2006). *New Review of Film and Television Studies*, 12(2), 156–177.

Martínez-Graña, A. M., Goy, J. L., & Cimarra, C. A. (2013). A virtual tour of geological heritage: Valourising geodiversity using Google Earth and QR code. *Computers & Geosciences*, 61, 83–93.

Massey, D. B., & Jess, P. M. (1995). *A place in the world? Places, cultures and globalization.* Oxford, UK: The Open University.

Mazumdar, S., & Mazumdar, S. (1993). Sacred space and place attachment. *Journal of Environmental Psychology*, 13(3), 231–242.

Mazumdar, S., & Mazumdar, S. (2004). Religion and place attachment: A study of sacred places. *Journal of Environmental Psychology*, 24(3), 385–397.

McAnany, P. A. (2016). *Maya cultural heritage: How archaeologists and indigenous communities engage the past.* Lanham, MD: Rowman & Littlefield.

Mccrone, J. (2019). Christchurch and its search for a post-recovery sense of purpose. *Stuff*, October 26. Retrieved from https://www.stuff.co.nz/the-press/business/the-rebuild/116782765/christchurch-and-its-search-for-a-postrecovery-sense-of-purpose

McIntosh, A. J., Zygadlo, F. K., & Matunga, H. (2004). Rethinking Maori tourism. *Asia Pacific Journal of Tourism Research*, 9(4), 331–352.

McIvor, D., Paton, D., & Johnston, D. (2009). Modelling community preparation for natural hazards: Understanding hazard cognitions. *Journal of Pacific Rim Psychology*, 3(2), 39–46.

McKee, R., & Fryer, B. (2003). Storytelling that moves people. *Harvard Business Review*, 81(6), 51–55.

McKinzie, A. E. (2019). You don't miss it 'til it's gone: Insecurity, place, and the social construction of the environment. *Environmental Sociology*, 5(3), 232–242.

McLeod, S. (2020). Maslow's hierarchy of needs. *Simply Psychology*. Retrieved from https://www.simplypsychology.org/maslow.html

Medina, L. K. (2003). Commoditizing culture: Tourism and Maya identity. *Annals of Tourism Research*, 30(2), 353–368.

Mei, X. Y., Hågensen, A. M. S., & Kristiansen, H. S. (2020). Storytelling through experiencescape: Creating unique stories and extraordinary experiences in farm tourism. *Tourism and Hospitality Research*, 20(1), 93–104.

Meinig, D. W. (1979). *The interpretation of ordinary landscapes: Geographical essays*. Oxford, UK: Oxford University Press.

Meloni, A., Fornara, F., & Carrus, G. (2019). Predicting pro-environmental behaviors in the urban context: The direct or moderated effect of urban stress, city identity, and worldviews. *Cities*, 88, 83–90.

Meloni, G., & Swinnen, J. (2018). Trade and terroir. The political economy of the world's first geographical indications. *Food Policy*, 81, 1–20.

Miceli, R., Sotgiu, I., & Settanni, M. (2008). Disaster preparedness and perception of flood risk: A study in an alpine valley in Italy. *Journal of Environmental Psychology*, 28(2), 164–173.

Milgram, P., & Kishino, F. (1994). A taxonomy of mixed reality visual displays. *IEICE Transactions on Information and Systems*, 77(12), 1321–1329.

Milligan, M. J. (1998). Interactional past and potential: The social construction of place attachment. *Symbolic Interaction*, 21(1), 1–33.

Ministry of Business, Innovation, and Employment and Department of Conservation (2019). *New Zealand – Aotearoa Government tourism strategy*. [pdf]. Retrieved from https://www.mbie.govt.nz/dmsdocument/5482-2019-new-zealand-aotearoa-government-tourism-strategy-pdf

Ministry of Business, Innovation & Employment (2020). *Māori in the Labour Market – June 2020 Quarter (unadjusted)*. Wellington: Ministry of Business, Innovation & Employment. Retrieved from: https://www.mbie.govt.nz/dmsdocument/11844-maori-in-the-labour-market-june-2020-quarter-pdf

Mishra, S., Mazumdar, S., & Suar, D. (2010). Place attachment and flood preparedness. *Journal of Environmental Psychology*, 30(2), 187–197.

Mitchell, K. (1997). Transnational discourse: Bringing geography back in. *Antipode*, 29(2), 101–114.

Montejo, V. (2004). Angering the ancestors: Transnationalism and economic transformation of Maya communities in western Guatemala. In J. M. Watanabe & E. F. Fischer (Eds.), *Pluralizing ethnography: Comparison and representation in Maya cultures, histories, and identities* (pp. 231–256). Santa Fe, NM: School of American Research Press.

Moore, R., Hankinson Gathman, E., & Ducheneaut, N. (2009). From 3D space to third place: The social life of small virtual spaces. *Human Organization*, 68(2), 230–240.

Moore, R. C. (1986). *Childhood's domain: Play and place in child development*. Berkeley, CA: MIG Publications.

Morgan, N., Pritchard, A., & Pride, R. (Eds.). (2011). *Destination brands: Managing place reputation* (3rd ed.). Oxford: Elsevier.

Myat Thu, P. (2020). Journeys to Knua: Displacement, return and translocality in Timor-Leste. *Mobilities*. https://doi.org/10.1080/17450101.2020.1742387

Müller, D. K., & Hall, C. M. (2003). Second homes and regional population distribution: On administrative practices and failures in Sweden. *Espace Population Societes*, 2003(2), 251–261.

Nabhan, G. P., & St. Antoine, S. (1993). The loss of floral and faunal story: The extinction of experience. In S. R. Kellert & E. O. Wilson (Eds.), *The biophilia hypothesis* (pp. 229–250). Washington, DC: Island Press.

Nagar, P. K., Gargava, P., Shukla, V. K., Sharma, M., Pathak, A. K., & Singh, D. (2020). Multi-pollutant air quality analyses and apportionment of sources in three particle size categories at Taj Mahal, Agra. *Atmospheric Pollution Research*. https://doi.org/10.1016/j.apr.2020.09.001.

Nagel, C. R., & Staeheli, L. A. (2004). Citizenship, identity and transnational migration: Arab immigrants to the United States. *Space and Polity*, 8(1), 3–23.

Nanson, A. (2011). *Words of re-enchantment: Writings on storytelling, myth, and ecological desire*. Bath, UK: Awen Publications.

Nash, C. (2002). Genealogical identities. *Environment and planning D: Society and Space*, 20(1), 27–52.

Neill, L., & Waring, M. (2019). Signifier of Kiwi identity. *Public History Review*, 26, 38–64.

New Zealand Māori Tourism Society. (2020). *Who is NZ Māori tourism*. Retrieved from https://maoritourism.co.nz/about/who-is-nz-maori-tourism

New Zealand Ministry for Culture and Heritage. (n.d.). *Scottish settlers arrive in Otago*. Retrieved from https://nzhistory.govt.nz/scottish-settlers-arrive-otago

Ngāi Tahu Tourism. (2020). *About us. The investment in tourism has helped Ngāi Tahu reconnect to areas of significant historical importance*. Retrieved from https://www.ngaitahutourism.co.nz/#about

Nicholson, R. (2020). Patty Jenkins is right: Nothing can replace the joy of watching a film on the big screen. *The Guardian*, 10 October. Retrieved from https://www.theguardian.com/commentisfree/2020/oct/10/patty-jenkins-is-right-nothing-can-replace-the-joy-of-watching-a-film-on-the-big-screen

Nicolosi, E., & Corbett, J. B. (2018). Engagement with climate change and the environment: A review of the role of relationships to place. *Local Environment*, 23(1), 77–99.

Nielsen, J. K. (2017). Museum communication and storytelling: Articulating understandings within the museum structure. *Museum Management and Curatorship*, 32(5), 440–455.

Nielsen-Pincus, M., Hall, T., Force, J. E., & Wulfhorst, J. D. (2010). Sociodemographic effects on place bonding. *Journal of Environmental Psychology*, 30(4), 443–454.

Nilsson, M., & Tesfahuney, M. (2016). Performing the "post-secular" in Santiago de Compostela. *Annals of Tourism Research*, 57, 18–30.

Nishiyama, K. (1996). *Welcoming the Japanese visitor: Insights, tips, tactics*. Honolulu, HI: University of Hawaii Press.

Nitsche, M. (2008). *Video game spaces: Image, play, and structure in 3D worlds*. Cambridge, MA: MIT Press.

Norman, A., & Pokorny, J. J. (2017). Meditation retreats: Spiritual tourism well-being interventions. *Tourism Management Perspectives*, 24, 201–207.

Norris, F. H., Stevens, S. P., Pfefferbaum, B., Wyche, K. F., & Pfefferbaum, R. L. (2008). Community resilience as a metaphor, theory, set of capacities, and strategy for disaster readiness. *American Journal of Community Psychology*, 41(1–2), 127–150.

Nouza, M., Ólafsdóttir, R., & Sæþórsdóttir, A. D. (2018). Motives and behaviour of second home owners in Iceland reflected by place attachment. *Current Issues in Tourism*, 21(2), 225–242.

O'Brien, D. J., Hassinger, E. W., & Dershem, L. (1994). Community attachment and depression among residents in two rural Midwestern communities. *Rural Sociology*, 59(2), 255–265.

Olsen, D. H., & Timothy, D. J. (2006). Tourism and religious journeys. In D. Timothy & D. Olsen (Eds.), *Tourism, religion and spiritual journeys* (pp. 17–38). London: Routledge.

Olsen, D. H., & Wilkinson, G. (2016). Are fast pilgrims true pilgrims? The Shikoku pilgrimage. *Annals of Tourism Research*, 61, 228–230.

Olsen, K. (2008). The Maori of tourist brochures representing indigenousness. *Journal of Tourism and Cultural Change*, 6(3), 161–184.

O'Neill, M., & Spybey, T. (2003). Global refugees, exile, displacement and belonging. *Sociology*, 37(1), 7–12.

Oppenheim, R. (1968). Māori culture. In A. McLeod (Ed.), *The pattern of New Zealand culture* (pp. 277–296). New York, NY: Cornell University Press.

Ostrow, A. P. (2002). Dual resident voting: Traditional disenfranchisement and prospects for change. *Columbia Law Review*, 102(7), 1954–1991.

Pacione, M. (2001). *Urban geography: A global perspective*. London: Routledge.

Pai, H.-H. (2020). The coronavirus crisis has exposed China's long history of racism. *The Guardian*, 25 April, https://www.theguardian.com/commentisfree/2020/apr/25/coronavirus-exposed-china-history-racism-africans-guangzhou.

Paiva, D. (2015). Experiencing virtual places: Insights on the geographies of sim racing. *Journal of Cultural Geography*, 32(2), 145–168.

Pantelidis, C. (2019). Exploring VR experiences of tourists' attachment to a rural destination. *International Journal of Technology Marketing*, 13(3–4), 376–400.

Pares, M., Marti-Costa, M., & Blanco, I. (2014). Geographies of governance: How place matters in urban regeneration policies. *Urban Studies*, 51(15), 3250–3267.

Paton, D., Smith, L., & Violanti, J. (2000). Disaster response: Risk, vulnerability and resilience. *Disaster Prevention and Management: An International Journal*, 9(3), 173–180.

Pearce, D. G. (1980). Tourism and regional development a genetic approach. *Annals of Tourism Research*, 7(1), 69–82.

Peters, G. L. (1997). *American winescapes: The cultural landscapes of America's wine country*. Boulder, CO: Westview Press.

Pei, Q., Zhang, D. D., & Lee, H. F. (2016). Contextualizing human migration in different agro-ecological zones in ancient China. *Quaternary International*, 426, 65–74.

Pelliccia, A. (2018). In the family home: Roots tourism among Greek second generation in Italy. *Current Issues in Tourism*, 21(18), 2108–2123.

Pera, R. (2017). Empowering the new traveller: Storytelling as a co-creative behaviour in tourism. *Current Issues in Tourism*, 20(4), 331–338.

Pera, R., & Viglia, G. (2016). Exploring how video digital storytelling builds relationship experiences. *Psychology & Marketing*, 33(12), 1142–1150.

Peters, F. E. (1996). *The Hajj: The Muslim pilgrimage to Mecca and the holy places*. Princeton, NJ: Princeton University Press.

Piper, N. (1998). *Racism, nationalism and citizenship: A comparative analysis of Britain and Germany*. Aldershot, UK: Ashgate.

Pletinckx, D., Silberman, N., & Callebaut, D. (2003). Heritage presentation through interactive storytelling: A new multimedia database approach. *The Journal of Visualization and Computer Animation*, 14(4), 225–231.

Plunkett, D. (2011). On place attachments in virtual worlds. *World Leisure Journal*, 53(3), 168–178.

Pocock, D. (1992). Catherine Cookson country: Tourist expectation and experience. *Geography*, 77(3), 236–243.

Poria, Y., Butler, R., & Airey, D. (2003). Tourism, religion and religiosity: A holy mess. *Current Issues in Tourism*, 6(4), 340–363.

Porter, J. E. (2004). Pilgrimage and the IDIC ethic: Exploring Star Trek convention attendance as pilgrimage. In E. Badone & S. R. Roseman (Eds.), *Intersecting journeys: The anthropology of pilgrimage and tourism* (pp. 160–179). Champaign, IL: University of Illinois Press.

Porter, J. E., & McLaren, D. L. (Eds.). (1999). *Star Trek and sacred ground: Explorations of Star Trek, religion, and American culture*. Albany, NY: State University of New York Press.

Prayag, G., & Lee, C. (2019). Tourist motivation and place attachment: The mediating effects of service interactions with hotel employees. *Journal of Travel & Tourism Marketing*, 36(1), 90–106.

Prayag, G., Mura, P., Hall, M., & Fontaine, J. (2015). Drug or spirituality seekers? Consuming ayahuasca. *Annals of Tourism Research*, 52(C), 175–177.

Prayag, G., Mura, P., Hall, C. M., & Fontaine, J. (2016). Spirituality, drugs, and tourism: Tourists' and shamans' experiences of ayahuasca in Iquitos, Peru. *Tourism Recreation Research*, 41(3), 314–325.

Prestes, P. (1995). Postmodern tourism: The Santa Claus industry. *Annals of Tourism Research*, 22(1), 1–15.

Pretty, G. H., Chipuer, H. M., & Bramston, P. (2003). Sense of place amongst adolescents and adults in two rural Australian towns: The discriminating features of place attachment, sense of community and place dependence in relation to place identity. *Journal of Environmental Psychology*, 23(3), 273–287.

Project for Public Spaces. (2007). *What is placemaking?*, Retrieved from https://www.pps.org/article/what-is-placemaking

Project for Public Spaces. (1975–2018). *What makes a successful place?*, Retrieved from https://www.pps.org/article/grplacefeat

Proshansky, H. M. (1978). The city and self-identity. *Environment and Behavior*, 10(2), 147–169.

Proshansky, H. M., Fabian, A. K., & Kaminoff, R. (1983). Place-identity: Physical world socialization of the self. *Journal of Environmental Psychology*, 3(1), 57–83.

Provenzano, D. (2020). The migration–tourism nexus in the EU28. *Tourism Economics*. https://doi.org/10.1177%2F1354816620909994

Puddifoot, J. E. (1995). Dimensions of community identity. *Journal of Community & Applied Social Psychology*, 5(5), 357–370.

Puriri, A., & McIntosh, A. (2019). A cultural framework for Māori tourism: Values and processes of a Whānau tourism business development. *Journal of the Royal Society of New Zealand*, 49(sup1), 89–103.

Qian, J., & Zhu, H. (2014). Chinese urban migrants' sense of place: Emotional attachment, identity formation, and place dependence in the city and community of Guangzhou. *Asia Pacific Viewpoint*, 55(1), 81–101.

Qing, P., Xi, A., & Hu, W. (2015). Self-consumption, gifting, and Chinese wine consumers. *Canadian Journal of Agricultural Economics/Revue canadienne d'agroeconomie*, 63(4), 601–620.

Qu, H., & Lam, S. (1997). A travel demand model for Mainland Chinese tourists to Hong Kong. *Tourism Management*, 18(8), 593–597.

Quach, G. (2020). Back to where you came from: How Vietnam drama Monsoon ignites the battle for belonging. *The Guardian*, 29 September, https://www.theguardian.com/film/2020/sep/29/vietnam-drama-monsoon-battle-for-belonging-hong-khaou

Raffaetà, R., & Duff, C. (2013). Putting belonging into place: Place experience and sense of belonging among Ecuadorian migrants in an Italian Alpine region. *City & Society*, 25(3), 328–347.

Ralph, D. (2009). 'Home is where the heart is'? Understandings of 'home' among Irish-born return migrants from the United States. *Irish Studies Review*, 17(2), 183–200.

Ramsay, G. (2003). Cherbourg's Chinatown: Creating an identity of place on an Australian Aboriginal settlement. *Journal of Historical Geography*, 29(1), 109–122.

Raviv, A., Keinan, G., Abazon, Y., & Raviv, A. (1990). Moving as a stressful life event for adolescents. *Journal of Community Psychology*, 18(2), 130–140.

Reef, C. (2006). *E.E. Cummings a poet's life*. New York, NY: Clarion Books.

Reijnders, S. (2016). Stories that move: Fiction, imagination, tourism. *European Journal of Cultural Studies*, 19(6), 672–689.

Relph, E. (1976). *Place and placelessness*. London, UK: Pion.

Relph, E. (2002). Place. In I. Douglas, R. Huggett, & M. Robinson (Eds.), *Companion Encyclopedia of geography: The environment and humankind* (pp. 906–922). London, UK: Routledge.

Relph, E. (2007a). On the identity of places. In M. Carmona & S. Tiesdell (Eds.), *Urban design reader* (pp. 103–107). Oxford, UK: Architectural Press.

Relph, E. (2007b). Spirit of place and sense of place in virtual realities. *Techné: Research in Philosophy and Technology*, 10(3), 17–25.

Rickly-Boyd, J. M. (2013). Existential authenticity: Place matters. *Tourism Geographies*, 15(4), 680–686.

Rinne, J., Kietäväinen, A., Tuulentie, S., & Paloniemi, R. (2014). Governing second homes: A study of policy coherence of four policy areas in Finland, *Tourism Review International*, 18(3): 223–236.

Rinschede, G., & Bhardwaj, S. M. (Eds.). (1990). *Pilgrimage in the United States*. Berlin: Dietrich Reimer Verlag.

Rowles, G. D. (1983). Place and personal identity in old age: Observations from Appalachia. *Journal of Environmental Psychology*, 3(4), 299–313.

Ruback, R. B., Pandey, J., & Kohli, N. (2008). Evaluations of a sacred place: Role and religious belief at the Magh Mela. *Journal of Environmental Psychology*, 28(2), 174–184.

Ryan, C. (1997). Maori and tourism: A relationship of history, constitutions and rites. *Journal of Sustainable Tourism*, 5(4), 257–278.

Ryan, C. (2002). Tourism and cultural proximity: Examples from New Zealand. *Annals of Tourism Research*, 29(4), 952–971.

Ryan, C., & Crotts, J. (1997). Carving and tourism: A Maori perspective. *Annals of Tourism Research*, 24(4), 898–918.

Ryden, K. C. (1993). *Mapping the invisible landscape: Folklore, writing, and the sense of place*. Iowa City, IA: University of Iowa Press.

Sakakibara, C. (2008). "Our home is drowning": IÑupiat storytelling and climate change in Point Hope, Alaskalaska. *Geographical Review*, 98(4), 456–475.

Sandlos, J. (1998). The storied curriculum: Oral narrative, ethics, and environmental education. *The Journal of Environmental Education*, 30(1), 5–9.

Sanger, M. (1997). Sense of place and education. *The Journal of Environmental Education*, 29(1), 4–8.

Sæþórsdóttir, A. D., & Hall, C. M. (2020). Visitor satisfaction in wilderness in times of overtourism: A longitudinal study. *Journal of Sustainable Tourism*. https://doi.org/10.1080/09669582.2020.1817050.

Sæþórsdóttir, A. D., Hall, C. M., & Wendt, M. (2020). From boiling to frozen? The rise and fall of international tourism to Iceland in the era of overtourism. *Environments*, 7(8), 59.

Scannell, L., & Gifford, R. (2010). The relations between natural and civic place attachment and pro-environmental behavior. *Journal of Environmental Psychology*, 30(3), 289–297.

Seamon, D. (2014). Place attachment in phenomenology: The synergistic dynamism of place. In L. C. Manzo & P. Devine-Wright (Eds.), *Place attachment: Advances in theory, methods and application* (pp. 11–22). New York, NY: Routledge.

Scott, D., Gössling, S., & Hall, C. M. (2012). International tourism and climate change. *Wiley Interdisciplinary Reviews: Climate Change*, 3(3), 213–232.

Sequeira, L., & Morgado, L. (2013). Virtual archaeology in second life and opensimulator. *Journal of Virtual Worlds Research*, 6(1). https://www.learntechlib.org/p/178163/

Seraphin, H., Ivanov, S., Dosquet, F., & Bourliataux-Lajoinie, S. (2020). Archetypes of locals in destinations victim of overtourism. *Journal of Hospitality and Tourism Management*, 43, 283–288.

Sexton, J. D. (1999). *Mayan folktales: Folklore from Lake Atitlán, Guatemala*. Santa Fe, NM: University of New Mexico Press.

Seyfi, S., & Hall, C. M. (2020). Cultural heritage tourism in the MENA: Introduction and background. In C. M. Hall & S. Seyfi (Eds.), *Cultural and heritage tourism in the Middle East and North Africa* (pp. 1–33). Abingdon, UK: Routledge.

Shamai, S. (1991). Sense of place: An empirical measurement. *Geoforum*, 22(3), 347–358.

Shamai, S. (2018). Measuring negative sense of place: Israeli settlers' forced migration. *GeoJournal*, 83(6), 1349–1359.

Shamsuddin, S., & Ujang, N. (2008). Making places: The role of attachment in creating the sense of place for traditional streets in Malaysia. *Habitat International*, 32(3), 399–409.

Sharpley, R. (2009). Tourism, religion and spirituality. In T. Jamal & M. Robinson (Eds.), *The Sage handbook of tourism studies* (pp. 237–253). Thousand Oaks: Sage Publications.

Sharpley, R., & Jepson, D. (2011). Rural tourism: A spiritual experience?. *Annals of Tourism Research*, 38(1), 52–71.

Sheng, X. (2005). Chinese families. In B. N. Adams & J. Trost (Eds.), *Handbook of world families*, (pp. 99–128). Thousand Oaks, California: Sage.

Sherwood, S. (2020). Photographer assaulted in Christchurch's CBD. *Stuff*, May 1. Retrieved from https://www.stuff.co.nz/national/crime/121377383/photographer-assaulted-in-christchurchs-cbd

Shin, K. J., Nakakido, R., Horie, S., & Managi, S. (2016). The effects of community attachment and information seeking on displaced disaster victims' decision making. *PloS One*, 11(3), e0151928. https://doi.org/10.1371/journal.pone.0151928

Sigurdson, O. (2010). Beyond secularism? Towards a post-secular political theology. *Modern Theology*, 26(2), 177–196.

Silver, A., & Grek-Martin, J. (2015). "Now we understand what community really means": Reconceptualizing the role of sense of place in the disaster recovery process. *Journal of Environmental Psychology*, 42, 32–41.

Simpson, D. (2006). 'Da Vinci Code' helps Louvre break visitor record. *CABI*. Retrieved from https://www.cabi.org/leisuretourism/news/15085

Singh, N., & Lee, M. J. (2009). Exploring perceptions toward education in 3-D virtual environments: An introduction to "second life". *Journal of Teaching in Travel & Tourism*, 8(4), 315–327.

Singh, P., Charan, D., Kaur, M., Railoa, K., & Chand, R. (2020). Place attachment and cultural barriers to climate change induced relocation: Lessons from Vunisavisavi Village, Vanua Levu, Fiji. In W. L. Filho (Ed.), *Managing climate change adaptation in the Pacific Region* (pp. 27–43). Gland, Switzerland: Springer Nature.

Singh, R., & Sharma, B. S. (2012). Composition, seasonal variation, and sources of PM 10 from World Heritage site Taj Mahal, Agra. *Environmental Monitoring and Assessment*, 184(10), 5945–5956.

Singh, R. P. (2006). Pilgrimage in Hinduism: Historical context and modern perspectives. In D. J. Timothy & D. H. Olsen (Eds.), *Tourism, religion and spiritual journeys* (pp. 236–252). London: Routledge.

Skipper, J., & Davidson, S. R. (2018). The big house as home: Roots tourism and slavery in the USA. *International Journal of Tourism Anthropology*, 6(4), 390–410.

Skrbis, Z., Woodward, I., & Bean, C. (2014). Seeds of cosmopolitan future? Young people and their aspirations for future mobility. *Journal of Youth Studies*, 17(5), 614–625.

Slater, M., Sadagic, A., Usoh, M., & Schroeder, R. (2000). Small-group behavior in a virtual and real environment: A comparative study. *Presence: Teleoperators & Virtual Environments*, 9(1), 37–51.

Smith, J. W., Siderelis, C., & Moore, R. L. (2010). The effects of place attachment, hypothetical site modifications and use levels on recreation behavior. *Journal of Leisure Research*, 42(4), 621–640.

Snellman, R. L. (2006). Santa tourism superpower. Four days in Lapland can cost as much as a fortnight in the Caribbean, but it's worth it. *Helsingin Sanomat International Edition – Business and Finance*, 19 December. Retrieved from http://www.hs.fi/english/article/Santa%20Tourism%20Superpower/1135223794225

Song, F., & Cho, M. S. (2017). Geography of food consumption patterns between south and north China. *Foods*, 6(5), 34–46.

Song, M. (2003). *Choosing ethnic identity*. Cambridge, UK: Polity Press.

Speller, G., Lyons, E., & Twigger-Ross, C. (2002). A community in transition: The relationship between spatial change and identity processes. *Social Psychological Review*, 4(2), 39–58.

Stausberg, M. (2014). Religion and spirituality in tourism. In A. A. Lew, C. M. Hall & A. M. Williams (Eds.), *The Wiley Blackwell companion to tourism* (pp.349–360). Hoboken: Wiley.

Stedman, R. C. (2002). Toward a social psychology of place: Predicting behavior from place-based cognitions, attitude, and identity. *Environment and Behavior*, 34(5), 561–581.

Stedman, R. C. (2003). Sense of place and forest science: Toward a program of quantitative research. *Forest Science*, 49(6), 822–829.

Stedman, R. C. (2006). Understanding place attachment among second home owners. *American Behavioral Scientist*, 50(2), 187–205.

Stein, M. (2004). The critical period of disasters: Insights from sense-making and psychoanalytic theory. *Human Relations*, 57(10), 1243–1261.

Steinkuehler, C. A., & Williams, D. (2006). Where everybody knows your (screen) name: Online games as "third places". *Journal of Computer-Mediated Communication*, 11(4), 885–909.

Steuer, J. (1992). Defining virtual reality: Dimensions determining telepresence. *Journal of Communication*, 42(4), 73–93.

Strydom, W., Puren, K., & Drewes, E. (2018). Exploring theoretical trends in placemaking: Towards new perspectives in spatial planning. *Journal of Place Management and Development*, 11(2), 165–180.

Stokols, D., & Shumaker, S. (1981). People in places: A transactional view of settings. In J. H. Harvey (Ed.), *Cognition, social behavior, and the environment* (pp. 441–488). Hillsdale, NJ: Lawrence Erlbaum Associates.

Stokowski, P. A. (2002). Languages of place and discourses of power: Constructing new senses of place. *Journal of Leisure Research*, 34(4), 368–382.

Strom, E. (2017). How place matters: A view from the Sunbelt. *Urban Affairs Review*, 53(1), 197–209.

Suh, A., & Prophet, J. (2018). The state of immersive technology research: A literature analysis. *Computers in Human Behavior*, 86, 77–90.

Sun, J., Ling, L., & Huang, Z. J. (2020). Tourism migrant workers: The internal integration from urban to rural destinations. *Annals of Tourism Research*, 84. https://doi.org/10.1016/j.annals.2020.102972

Surovaya, E., Prayag, G., Yung, R., & Khoo-Lattimore, C. (2020). Telepresent or not? Virtual reality, service perceptions, emotions and post-consumption behaviors. *Anatolia*, https://doi.org/10.1080/13032917.2020.1808431

Tahana, N., Grant, K. T. O. K., Simmons, D. G., & Fairweather, J. R. (2000). *Tourism and Maori development in Rotorua*. Report No. 15. Tourism Recreation Research and Education Centre (T.R.R.E.C.), Lincoln University, New Zealand.

Tang, Z. (2017). Changes with Chinese characteristics: Rural clan culture, clan communities, and kinship relations during urbanization. In Z. Tang (Ed.), *China's urbanization and socioeconomic impact* (pp. 161–198). Springer, Singapore.

Taylor, S. R. (2018). *On being Maya and getting by: Heritage politics and community development in Yucatán*. Louisville, CO: University Press of Colorado

Te Ara – the Encyclopedia of New Zealand. (n.d.). *Te Ahukaramū Charles Royal, 'Whenua – how the land was shaped – The North and South islands'*. Retrieved from https://teara.govt.nz/en/whenua-how-the-land-was-shaped/page-2

Te Rito, J. S. (2007). Whakapapa: A framework for understanding identity. *MAI Review*, 1(3), 10–19.

Te Rūnanga o Ngāi Tahu. (1997). *Ngāi Tahu – the iwi*. Retrieved from https://ngaitahu.iwi.nz/ngai-tahu/

Tervo, K. (2008). The operational and regional vulnerability of winter tourism to climate variability and change: The case of the Finnish nature-based tourism entrepreneurs. *Scandinavian Journal of Hospitality and Tourism*, 8(4), 317–332.

Tervo-Kankare, K., Hall, C. M., & Saarinen, J. (2013). Christmas tourists' perceptions to climate change in Rovaniemi, Finland. *Tourism Geographies*, 15(2), 292–317.

Terzidou, M. (2020). Re-materialising the religious tourism experience: A post-human perspective. *Annals of Tourism Research*, 83. https://doi.org/10.1016/j.annals.2020.102924

Teye, V., Sirakaya, E., & Sönmez, S. F. (2002). Residents' attitudes toward tourism development. *Annals of Tourism Research*, 29(3), 668–688.

Thompson, J. P. (2017). Place matters, and so does race. *Urban Affairs Review*, 53(1), 210–218.

Thouvenot, C. (1978). Studies in food geography in France. *Social Science & Medicine. Part D: Medical Geography*, 12(1), 43–54.

Tiberghien, G., & Xie, P. F. (2018). The life cycle of authenticity: Neo-nomadic tourism culture in Kazakhstan. *Journal of Tourism and Cultural Change*, 16(3), 234–247.

Timothy, D. J. (2008). Genealogical mobility: Tourism and the search for a personal past. In D. J. Timothy & J. K. Guelke (Eds.), *Geography and genealogy: Locating personal pasts* (pp. 115–135). Aldershot, UK: Ashgate Publishing.

Timothy, D. J., & D. Olsen, D. (Eds.). (2006). *Tourism, religion and spiritual journeys*. New York: Routledge.

Ting-Toomey, S. (2015). Identity negotiation theory. *The International Encyclopedia of Interpersonal Communication*. https://doi.org/10.1002/9781118540190.wbeic129.

Tolkien, J. R. R. (2012). *The Lord of the Rings*. Boston, MA: Houghton Mifflin Harcourt.

Tombs, A. G., & McColl-Kennedy, J. R. (2003). Social-servicescape conceptual model. *Marketing Theory*, 3(4), 447–475.

Tombs, A. G., Finsterwalder, J., Chen, C., Prayag, G., & Hall, C. M. (2017). *The role of transitional servicescapes in maintaining attachment to place*. Paper presented at the *Academy of Marketing Science (AMS) World Marketing Congress (WMC)*, Christchurch 27 June – 1 July 2017.

Tonge, J., Ryan, M. M., Moore, S. A., & Beckley, L. E. (2015). The effect of place attachment on pro-environment behavioral intentions of visitors to coastal natural area tourist destinations. *Journal of Travel Research*, 54(6), 730–743.

Tourism Research Australia. (2020). *International tourism statistics*. Retrieved from http://tra.gov.au/research/International-tourism-statistics.html

Tsai, C. T. (2016). Memorable tourist experiences and place attachment when consuming local food. *International Journal of Tourism Research*, 18(6), 536–548.

Tuan, Y. F. (1972). Topophilia: Personal encounters with the landscape. In P. W. English & R. C. Mayfield (Eds.), *Man, space, and environment* (pp. 534–538). New York: Oxford University Press.

Tuan, Y. F. (1974). *Topophilia: A study of environmental attitudes, perceptions and values*. Englewood Cliffs, NJ: Prentice Hall.

Tuan, Y. F. (1977). *Space and place: The perspective of experience*. Minneapolis, MN: University of Minnesota Press.

Tuan, Y. F. (1979) Space and place: Humanistic perspective. In S. Gale & G. Olsson (Eds.), *Philosophy in geography* (pp. 387–427). Dordrecht: Springer.

Tuan, Y. F. (1980). Rootedness versus sense of place. *Landscape*, 24, 3–8.

Tuan, Y. F. (1991). A view of geography. *Geographical Review*, 81, 99–107.

Turner, P., & Turner, S. (2006). Place, sense of place, and presence. *Presence: Teleoperators & Virtual Environments*, 15(2), 204–217.

Turner, P., Turner, S., & Burrows, L. (2013). Creating a sense of place with a deliberately constrained virtual environment. *International Journal of Cognitive Performance Support*, 1(1), 54–68.

Turner, V. (1973). The center out there: Pilgrim's goal. *History of Religions*, 12(3), 191–230.

Tussyadiah, I. P., Wang, D., Jung, T. H., & tom Dieck, M. C. (2018). Virtual reality, presence, and attitude change: Empirical evidence from tourism. *Tourism Management*, 66, 140–154.

Twigger-Ross, C. L., Bonaiuto, M., & Breakwell, G. M. (2003). Identity theories and environmental psychology. In M. Bonnes, T. Lee, & M. Bonaiuto (Eds.), *Psychological theories for environmental issues* (pp. 203–234). Aldershot, UK: Ashgate.

Twigger-Ross, C. L., & Uzzell, D. L. (1996). Place and identity processes. *Journal of Environmental Psychology*, 16(3), 205–220.

Uzzell, D. L. (1996). Creating place identity through heritage interpretation. *International Journal of Heritage Studies*, 1(4), 219–228.

Vaka'uta, C. K. F. (2012). *Cyberspace, place, identity & relationships: Are we digitizing the Vā?* Panel presentation on *Culture, Education and ICT – Cultural Dynamics of the 21st Century, International Conference on ICT & Oceanian Cultures, University of the South Pacific*, Suva Fiji, 24 February 2012.

Valdés, M. J. (2004). Story-telling and cultural identity in Latin America. In I. Blayer & M. Anderson (Eds.), *Latin American narratives and cultural identity: Selected readings* (pp. 9–27). New York, NY: Peter Lang Publishing.

Venäläinen, A., Lehtonen, I., Laapas, M., Ruosteenoja, K., Tikkanen, O. P., Viiri, H., … Peltola, H. (2020). Climate change induces multiple risks to boreal forests and forestry in Finland: A literature review. *Global Change Biology*, 26(8), 4178–4196.

Vaske, J. J., & Kobrin, K. C. (2001). Place attachment and environmentally responsible behavior. *The Journal of Environmental Education*, 32(4), 16–21.

VFS Global. (n.d.). *After submission*. Retrieved from https://www.vfsglobal.com/Netherlands/UK/visa-services/after-submission.html

Vors, L. S., & Boyce, M. S. (2009). Global declines of caribou and reindeer. *Global Change Biology*, 15, 2626–2633.

Waiti, J. T. A., & Awatere, S. (2019). Kaihekengaru: Māori surfers' and a sense of place. *Journal of Coastal Research*, 87(SI), 35–43.

Walby, S. (1994). Is citizenship gendered? *Sociology*, 28(2), 379–395.

Wang, F. L. (2005). *Organizing through division and exclusion: China's hukou system*. Stanford, CA: Stanford University Press.

Wang, H., Yang, Z., Chen, L., Yang, J., & Li, R. (2010). Minority community participation in tourism: A case of Kanas Tuva villages in Xinjiang, China. *Tourism Management*, 31(6), 759–764.

Wang, S., & Chen, J. S. (2015). The influence of place identity on perceived tourism impacts. *Annals of Tourism Research*, 52, 16–28.

Wang, S., Wang, J., Li, J., & Zhou, K. (2020a). How and when does religiosity contribute to tourists' intention to behave pro-environmentally in hotels? *Journal of Sustainable Tourism*, 28(8), 1120–1137.

Wang, S. S., & Hsieh, C. T. (2020). Ubiquitous Pokémon Go: Human–environment relationships and the location-based augmented reality game. *Environment and Behavior*, 52(7), 695–725.

Wang, W. W., & Fan, C. C. (2006). Success or failure: Selectivity and reasons of return migration in Sichuan and Anhui, China. *Environment and Planning A*, 38(5), 939–958.

Wang, Y., Shen, H., & Ye, S. (2020). Being rational and emotional: An integrated model of residents' support of ethnic tourism development. *Journal of Hospitality and Tourism Management*, 44, 112–121.

Wasfi, R., Steinmetz-Wood, M., & Kestens, Y. (2017). Place matters: A longitudinal analysis measuring the association between neighbourhood walkability and walking by age group and population center size in Canada. *PLoS One*, 12(12), e0189472.

Watkin, R. (2018). Placing Canadian geographical indications on the map. *Intellectual Property Journal*, 30(2), 271–308.

Watson, J. L. (1982). Chinese kinship reconsidered: Anthropological perspectives on historical research. *China Quarterly*, 92, 589–622.

Weaver, D. B. (2013). Protected area visitor willingness to participate in site enhancement activities. *Journal of Travel Research*, 52(3), 377–391.

Weick, K. E. (1995). *Sensemaking in organizations*. Thousand Oaks, CA: Sage.

White, D. D., Virden, R. J., & Van Riper, C. J. (2008). Effects of place identity, place dependence, and experience-use history on perceptions of recreation impacts in a natural setting. *Environmental Management*, 42(4), 647–657.

White, M. (2019). Too many tourists: Should we limit visitor numbers to NZ? *North & South*, August 22. Retrieved from https://www.noted.co.nz/money/money-economy/nz-tourists-should-we-limit-number-visitors

Whitehead, G. (2017). The contradictory condition of 'homelessness' in the life of the transnational professional. *Culture and Organization*, 23(4), 306–324.

Whyte, W. H. (1988). *City: Rediscovering the center*. New York: Doubleday.

Whyte, W. H. (1996). The design of spaces. In R. T. Le Gates & F. Stout (Eds.), *The city reader* (2nd ed., pp. 587–595). New York: Routledge.

Wilkins, E. J., & de Urioste-Stone, S. (2018). Place attachment, recreational activities, and travel intent under changing climate conditions. *Journal of Sustainable Tourism*, 26(5), 798–811.

Williams, A. M., & Hall, C. M. (2000). Tourism and migration: New relationships between production and consumption. *Tourism Geographies*, 2(1), 5–27.

Williams, D. R., & Vaske, J. J. (2003). The measurement of place attachment: Validity and generalizability of a psychometric approach. *Forest Science*, 49(6), 830–840.

Williams, H. W. (1971). *A dictionary of the Maori language*. Wellington, NZ: Government Printer Books.

Williams, R. (2009). Place attachment, sense of belonging and identity in life history narratives of Iranian Baha'i refugees. *Baha'i Studies Review*, 15(1), 3–18.

Wilson, R. (1997). A sense of place. *Early Childhood Education Journal*, 24(3), 191–194.

Wirth, D. M., & Gamon, J. A. (1999). The art of situated narrative: A tool to teach environmental ethics. *Journal of Vocational Education Research*, 24(1), 45–61.

Woods, M. (2011). The local politics of the global countryside: Boosterism, aspirational ruralism and the contested reconstitution of Queenstown, New Zealand. *GeoJournal*, 76(4), 365–381.

Wu, Q. (2006). Commercialization of digital storytelling: An integrated approach for cultural tourism, the Beijing Olympics and wireless VAS. *International Journal of Cultural Studies*, 9(3), 383–394.

Wu, Z., & So, V. W. (2020). Ethnic and national sense of belonging in Canadian society. *International Migration*, 58(2), 233–254.

Xiong, Y., Zhang, Y., & Lee, T. J. (2020). The rural creative class: An analysis of in-migration tourism entrepreneurship. *International Journal of Tourism Research*, 22(1), 42–53.

Xue, L., Kerstetter, D., & Hunt, C. (2017). Tourism development and changing rural identity in China. *Annals of Tourism Research*, 66, 170–182.

Xygalatas, D., Kotherová, S., Maňo, P., Kundt, R., Cigán, J., Klocová, E. K., & Lang, M. (2018). Big gods in small places: The random allocation game in Mauritius. *Religion, Brain & Behavior*, 8(2), 243–261.

Xygalatas, D., Klocová, E. K., Cigán, J., Kundt, R., Maňo, P., Kotherová, S., Mitkidis, P., Wallot, S., & Kanovsky, M. (2016). Location, location, location: Effects of cross-religious primes on prosocial behavior. *The International Journal for the Psychology of Religion*, 26(4), 304–319.

Yang, L., & Wall, G. (2008). Ethnic tourism and entrepreneurship: Xishuangbanna, Yunnan, China. *Tourism Geographies*, 10(4), 522–544.

Yang, L., Wall, G., & Smith, S. L. (2008). Ethnic tourism development: Chinese government perspectives. *Annals of Tourism Research*, 35(3), 751–771.

Yankholmes, A., & McKercher, B. (2020). Migrant identity and subsequent travel behaviour. *International Journal of Tourism Research*, 22(5), 537–550.

Yeh, C.-H., Wang, Y.-S., Li, H.-T., & Lin, S.-Y. (2017). The effect of information presentation modes on tourists' responses in Internet marketing: The moderating role of emotions. *Journal of Travel & Tourism Marketing*, 34(8), 1018–1032.

Yi, X., Lin, V. S., Jin, W., & Luo, Q. (2017). The authenticity of heritage sites, tourists' quest for existential authenticity, and destination loyalty. *Journal of Travel Research*, 56(8), 1032–1048.

Young, M. (1999). The social construction of tourist places. *Australian Geographer*, 30(3), 373–389.

Yu, L. (1997). Travel between politically divided China and Taiwan. *Asia Pacific Journal of Tourism Research*, 2(1), 19–30.

Yu, M., Li, Z., Yu, Z., He, J., & Zhou, J. (2020). Communication related health crisis on social media: A case of COVID-19 outbreak. *Current Issues in Tourism*. https://doi.org/10.1080/13683500.2020.1752632.

Yuan, Y., & Manuel, J. I. (2018). The relationship between residential mobility and behavioral health service use in a national sample of adults with mental health and/or substance abuse problems. *Journal of Dual Diagnosis*, 14(4), 201–210.

Yung, R., & Khoo-Lattimore, C. (2019). New realities: A systematic literature review on virtual reality and augmented reality in tourism research. *Current Issues in Tourism*, 22(17), 2056–2081.

Yung, R., Khoo-Lattimore, C., Prayag, G., & Surovaya, E. (2020). Around the world in less than a day: Virtual reality, destination image and perceived destination choice risk in family tourism. *Tourism Recreation Research*. https://doi.org/10.1080/02508281.2020.1788351

Yuval-Davis, N. (1997). *Gender and nation*. London, UK: Sage Publications.

Zavar, E., Lavy, B. L., & Hagelman III, R. R. (2020). Chain tourism in post-disaster recovery. *Tourist Studies*, 20(4), 429–449.

Zetter, R., & Boano, C. (2010). Space and place after natural disasters and forced displacement. In G. Lizarralde, C. Johnson, & C. Davidson (Eds.), *Rebuilding after disasters: From emergency to sustainability*, (pp. 206–230). New York: Spoon Press.

Zhang, C., Perkis, A., & Arndt, S. (2017, May). Spatial immersion versus emotional immersion, which is more immersive? In *2017 Ninth International Conference on Quality of Multimedia Experience (QoMEX)* (pp. 1–6). Erfurt, Germany: IEEE.

Zhang, L. (2001). *Strangers in the city: Reconfigurations of space, power, and social networks within China's floating population*. Stanford, CA: Stanford University Press.

Zhang, Y., Zhang, H. L., Zhang, J., & Cheng, S. (2014). Predicting residents' pro-environmental behaviors at tourist sites: The role of awareness of disaster's consequences, values, and place attachment. *Journal of Environmental Psychology*, 40, 131–146.

Zheng, Y., Goh, E., & Wen, J. (2020). The effects of misleading media reports about COVID-19 on Chinese tourists' mental health: A perspective article. *Anatolia*, 31(2), 337–340.

Ziebertz, H. G., & Riegel, U. (2010). Europe: A post-secular society? *International Journal of Practical Theology*, 13(2), 293–308.

Zittoun, T. (2020). Nourish your imagination and you will be forever free. *Psyche*, 7 October, Retrieved from https://psyche.co/ideas/nourish-your-imagination-and-you-will-be-forever-free

Index

For Product Safety Concerns and Information please contact our EU representative GPSR@taylorandfrancis.com
Taylor & Francis Verlag GmbH, Kaufingerstraße 24, 80331 München, Germany

www.ingramcontent.com/pod-product-compliance
Lightning Source LLC
LaVergne TN
LVHW021247011026
849160LV00010BA/935

* 9 7 8 1 0 3 2 0 0 6 2 1 5 *